Exploring Sustainable
Biodiesel

Exploring Sustainable
Biodiesel

Amy Townsend
Billy Broas
Chelsea Jenkins
Kevin Ray

4880 Lower Valley Road Atglen, Pennsylvania 19310

Other Schiffer Books by Amy Townsend
Green Business: A Five-Part Model for Creating an Environmentally Responsible Company.
Other Schiffer Books on Related Subjects
Green Roofs: Ecological Design and Construction. Earth Pledge.

Designed by John P. Cheek
Type set in Bodoni Bd BT/Arrus BT

ISBN: 978-0-7643-2908-1
Printed in China

Schiffer Books are available at special discounts for bulk purchases for sales promotions or premiums. Special editions, including personalized covers, corporate imprints, and excerpts can be created in large quantities for special needs. For more information contact the publisher:

Published by Schiffer Publishing Ltd.
4880 Lower Valley Road
Atglen, PA 19310
Phone: (610) 593-1777; Fax: (610) 593-2002
E-mail: Info@schifferbooks.com

For the largest selection of fine reference books on this and related subjects, please visit our web site at **www.schifferbooks.com**
We are always looking for people to write books on new and related subjects. If you have an idea for a book please contact us at the above address.

This book may be purchased from the publisher.
Include $3.95 for shipping.
Please try your bookstore first.
You may write for a free catalog.

In Europe, Schiffer books are distributed by
Bushwood Books
6 Marksbury Ave.
Kew Gardens
Surrey TW9 4JF England
Phone: 44 (0) 20 8392-8585; Fax: 44 (0) 20 8392-9876
E-mail: info@bushwoodbooks.co.uk
Website: www.bushwoodbooks.co.uk
Free postage in the U.K., Europe; air mail at cost.

Printed on 100% post consumer recycled paper

Contents

Acknowledgments

Collectively, we would like to thank and acknowledge the efforts of the many individuals who have helped us in the research and production of this book. Thanks go to our editor Tina Skinner, Jeff Snyder, and the team at Schiffer Books for their keen interest in biodiesel and sustainability. In addition, we would like to offer our deepest appreciation to our families and friends for their unwavering support.

We also want to extend a heartfelt thank you to the very busy biodiesel folks who took the time to review this book in its various drafts. We are deeply grateful to Sarah Hill and Matt Steiman, each of whom reviewed one or more chapters and provided invaluable feedback.

Also, we would like to thank Kumar Plocher, Ian Heatwole, Cristina Siegel, Mac Minaudo, Robert Miller, Tom Leue, Nadia Adawi, Matt Steiman, and Jeremy Ferrell for providing case studies of their experiences working in the biodiesel industry.

Many thanks to the Virginia Department of Environmental Quality and the New Hampshire Department of Environmental Services, both of which provided vital regulatory information to us.

Thank you to Scott Burkholder, Lyle Estill, Michael Guymon, Matt Hollander, Jack Martin, and Jay Nance, who provided helpful information and suggestions to us during the development of this book. Any shortcomings or errors are fully our own.

While we worked on this project, we met and spoke with some truly amazing people who are committed to both raising environmental awareness and implementing alternatives to decrease our dependence on fossil fuels. Some described their own personal biodiesel mission and experiences while others provided guidance as we compiled this work. There are countless individuals to thank, including all of the biodiesel producers, resellers, and company founders who have been willing to share their knowledge of biodiesel. We would like to express deep gratitude and appreciation for the difficult work you perform on a daily basis.

Preface

We met through James Madison University's College of Integrated Science and Technology and had worked together briefly on biodiesel research projects overseas. When the opportunity to write a book about biodiesel arose from Schiffer Publishing, none of us hesitated long before agreeing to co-author this book.

We wrote *Exploring Sustainable Biodiesel* to draw more attention to the environmental issues surrounding biodiesel. We believe that biodiesel can play an important role as a cleaner, environmentally preferable fuel – if it is made and used responsibly. We hope that this book inspires producers to make more sustainable biodiesel and consumers to demand greener sources of biodiesel and other potentially renewable fuels.

Chapter 1. Introduction

This chapter discusses the growing interest in biodiesel. It explains what biodiesel is and how it is made. Then, it examines some of the benefits attributed to biodiesel and explores the need for sustainable biodiesel. Finally, it discusses the purpose of this book.

The Biodiesel Boom

There is a growing interest in biodiesel, and this is occurring for many reasons. Concern over climate change, environmental degradation, asthma and other respiratory illnesses, rising fuel prices, the health of local economies, and national security have inspired governments, individuals, and companies to search actively for more human- and environment-friendly fuels. Though many alternative energy sources are being developed and used, this book focuses on one – biodiesel. Biodiesel is one of several fuels that – together with telecommuting, reduced fuel use, greater fuel efficiency, and the development of effective mass transit systems – can end our dependence on petroleum to power our economy and our lives.

According to Biofuel Market Worldwide, "Growing at the rate of more than 30% from the year 2006, world Biodiesel production is likely to touch the mark of 12 Billion Liter [sic] by the end of 2010" (RNCOS, 2006). Biodiesel sales have risen markedly, and biodiesel is becoming an increasingly popular source of fuel or fuel additive in the US and abroad.

With certain caveats, biodiesel is one of the most promising near-term replacements for oil because of its easy integration into the current fuel structure. For example, it can be transported easily using existing rail and trucking infrastructure; it can be stored in existing tanks and dispersed through existing fuel pumps at gas stations; and it can be used in existing diesel engines with little or no modification. Biodiesel can be blended with petroleum diesel and used as a fuel additive due to its beneficial effects on lubricity (ability to lubricate). Moreover, if it is made and used sustainably, it can be a renewable resource and burns cleaner than either petroleum diesel fuel or gasoline. *We suggest that sustainable biodiesel does not diminish the integrity or resilience of any ecosystem, human community, or human economy during any stage of its life cycle.*

Biodiesel is a renewable fuel that can be far superior to petroleum diesel in many respects. Not only has biodiesel been used successfully in cars, trucks, boats, and other vehicles, but it also has been used successfully as a home heating oil substitute.

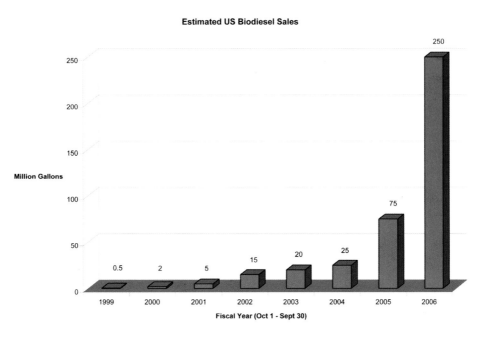

Estimated US biodiesel sales – 1999-2006. *Courtesy of National Biodiesel Board 2007g.*

What Is Biodiesel?

Biodiesel can be defined loosely as a vegetable oil-based or animal fat-based fuel, which can be burned in a diesel engine, for home heat, and other applications. However, it is important to state that biodiesel is not simply vegetable oil or animal fat. Although Rudolf Diesel designed the diesel engine to run on peanut oil, biodiesel is more than that. Biodiesel is a fuel made from fats, which, when mixed with other ingredients, undergo a chemical reaction that results in biodiesel. In its pure form (B100), it contains no petroleum products and can be a renewable energy source.

The National Biodiesel Board provides a more technical definition for biodiesel: "a fuel comprised of mono-alkyl esters of long chain fatty acids derived from vegetable oils or animal fats, designated B100, and meeting the requirements of ASTM D 6751" (NBB 2007i).

The properties of biodiesel are so similar to the properties of petroleum diesel that the two can be blended at any level. Blends of biodiesel and petroleum diesel commonly are referred as B"X," with X signifying the percentage of biodiesel in the blend. The most common blends are B2 and B20. Because of its solvent capabilities, vehicle owners who use a high blend of biodiesel should ensure that all gaskets, hoses, and o-rings are made of synthetic materials rather than rubber, which can break down over time. Moreover, biodiesel effectively cleans the fuel lines; as a result, the fuel filter can become clogged with debris and should be changed as needed.

Biodiesel is becoming a popular replacement for petroleum diesel fuel as it becomes more widely available and consumers learn about its benefits. In 1999, only 500,000 gallons of biodiesel were produced commercially in the United States (NBB 2007k). In 2006, about 250 million gallons of biodiesel were sold in the United States (NBB 2007k).

How Biodiesel Is Made

Biodiesel is the product of transesterification, a chemical reaction that is created using three ingredients – fat, alcohol, and a catalyst. Methanol is the alcohol, and lye is the catalyst that often are used (see the figure on the following page). Biodiesel is made at a low temperature and pressure, requiring little energy input relative to petroleum diesel. The highest temperature required is 150 degrees Fahrenheit, and the highest pressure required is 20psi. (For more information on biodiesel production, refer to www.biodiesel.org.)

Glycerin, which many are familiar with as an ingredient in some soaps, is a byproduct of the biodiesel production process. During biodiesel manufacturing, the glycerin must be separated from the oil. Then, the mixture is washed and purified until all traces of glycerin, water, and methanol are gone. Although the glycerin can be used to make soap, it must be purified since it contains the methanol used in biodiesel production. The figure below illustrates the biodiesel production process.

Frequently, biodiesel is blended with petroleum diesel due to several factors, which can include cost, lubricity, and engine warranty issues. Biodiesel made for sale must adhere to the American Society for Testing and Materials (ASTM) standard titled ASTM D 6751 and any changes to that standard, which is being updated continuously. Under this standard, commercially produced biodiesel must be tested according to several criteria, which are discussed in the chapter on biodiesel quality.

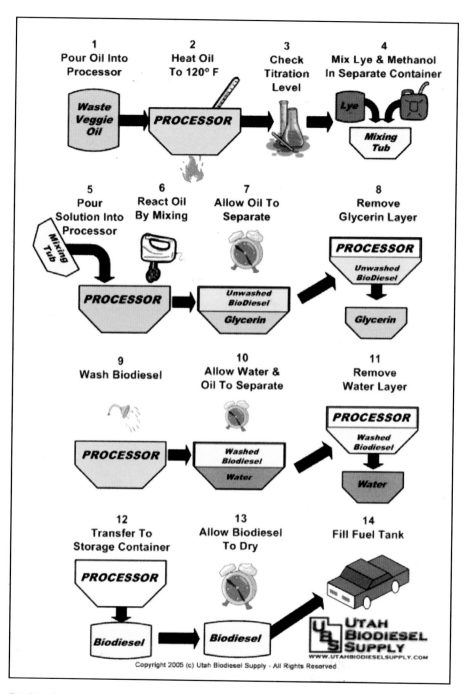

Biodiesel production process. *Courtesy of Utah Biodiesel Supply, http://www.utahbiodieselsupply.com/makingbiodiesel.php.*

Current State of Petroleum

Global oil reserves are being depleted at unprecedented rates, and the cost of getting the remaining oil will become increasingly expensive over time regardless of the new technologies developed to extract that oil. The resulting loss of oil would be a blow to the U.S. economy, which uses more oil than any other country in the world – a total of 20.7 million barrels of oil every day in 2004 (US Department of Energy, Energy Information Administration 2004). This was more than triple the amount burned by the world's second largest consumer of oil (US DOE EIA 2004e). About 60% of this oil is imported with about half of it coming from OPEC members (British Petroleum Co. 2007).

Oil has several uses, not the least of which is in transportation. In fact, the U.S. transportation sector accounts for two-thirds of the country's oil use, burning almost 700 million gallons of oil per annum. That is an average of 15.58 million barrels of oil per day (US DOE EIA 2005) or 2.3 gallons per day for every man, woman, and child (U.S. Census Bureau 2007).

More than 90% of the world's vehicles are powered by petroleum products (US DOE EIA 2004), and over 95% of the vehicles in the United States are powered by fossil fuels (Bush 2006). In 2006, alone, transportation used 28.4 quadrillion BTUs, which was 28.5% of the total energy use of all sectors (US DOE, March 2007). With transportation-related energy consumption projected to grow at an annual rate of 1.4 percent (US DOE, February 2007), on average, we need to either find more oil, curb our hunger for energy, or establish a much more renewably based and diverse transportation infrastructure. This first option is not very viable given the impact that burning oil has on the world's climate.

Kenneth Deffeyes, a geologist and author of *Beyond Oil*, wrote that by 2019, production will be down to 90 percent of the peak level (Deffeyes

2005). Others, too, have tried to predict when oil will peak. The International Energy Agency and Energy Information Agency have predicted that the world's oil production will peak around the middle of the twenty-first century at today's consumption rates (US DOE EIA 2004f). Of course, these numbers do not take into account the booming economies and related energy use of China or India (Shipper et al. 1997).

In its Statistical Review of World Energy, British Petroleum (2007) reported that there are about 1,208.2 billion barrels of proven oil reserves left that are accessible for drilling. The *Oil and Gas Journal*, World Energy, and the OPEC Secretariat all have developed similar statistics (Deffeyes 2001; World Oil 2005; OPEC 2005). If fossil fuel consumption were to remain at today's levels, this means that the world's oil reserves would be depleted in the next forty years. This provides us with just a few decades in which to find another, renewable source of energy.

What is the difference between per capita peak and peak level of oil? Per capita oil is reported on a country by country basis. Because few countries depend only on their own oil resources, analysts generally discuss the worldwide peak of oil reserves.

Why do we even care about peak oil production when it means that we have only consumed about half of the world's oil resource? The world can develop alternative sources of energy while using the other half of the oil – right? Of course, alternative energy sources already are being developed and used. However, when we hit that peak amount of oil, production will decline – slowly at first and, then, more rapidly. Production will not be able to meet demand, and the price of oil will escalate rapidly and continuously. If you were one of many who, following the rise in oil prices after Hurricane Katrina in 2005, believed that $3 per gallon of gas was outrageous, imagine a day in which you will view $8 per gallon as being reasonable or, even, inexpensive.

Unfortunately, while there are many alternative energy technologies, most are far from widespread in terms of market penetration. Regardless of whether peak oil occurs 10, 20, 30, or even 200 years from now, its time as the primary fuel source for human industry and activity is waning. This is true for several reasons. For example, with concerns over carbon dioxide emissions resulting from fossil fuel combustion and their impacts on the global climate, some climate scientists have indicated that we might not have a few decades during which to move to alternative energy sources.

Violence and instability in the Middle East, in which more than 60% of the world's oil is located (British Petroleum Co. 2005), also are a cause for concern. About 80% of the world's oil is controlled by OPEC's eleven members – Algeria, Indonesia, Iran, Iraq, Kuwait, Libya, Nigeria, Qatar, Saudi Arabia, the United Arab Emirates, and Venezuela (OPEC 2006).

The burgeoning human population and related natural resource consumption are creating an increased demand for the world's finite fossil fuel reserves. Much of the industrialized world's food production and distribution also relies heavily on fossil fuels for everything from farm equipment to herbicides, pesticides, and fertilizers.

The Benefits of Biodiesel

Biodiesel Emissions Are Healthier for the Environment and You

The transportation sector accounts for nearly one-third of US greenhouse gas emissions with projections rising to thirty-six percent by 2020 (Greene and Schafer 2003). Although behavioral changes could help to reduce the negative environmental impacts associated with this sector, the development of greener, alternative fuels is an important component of reducing the transportation sector's dependence on oil.

Biodiesel, one of these alternative fuels, offers many benefits. For instance, it is non-toxic, biodegradable, and free of sulfurs, aromatics, and other harmful substances that are associated with petroleum diesel. Studies by the US Environmental Protection Agency indicate that, compared to petroleum diesel fuel, biodiesel emits far fewer particulates, hydrocarbons, cancer-causing agents, and sulphur oxides. In fact, all of biodiesel's emissions are reduced with the possible exception of nitrogen oxides (NOx), which will be discussed later.

Biodiesel was the first alternative fuel to finish the US Environmental Protection Agency's Tier I and Tier II health effects testing requirements, which were established by the 1990 Clean Air Act Amendments (US Government Printing Office 2001). Using its health effects testing requirements, the Agency compared biodiesel to petroleum diesel fuel to determine the environmental, including human, health of biodiesel. Among other things, the study concluded that biodiesel is as biodegradable as sugar, ten times less toxic than table salt, and has lower emissions than petroleum diesel fuel. Some of the other results are outlined below.

- "The overall *ozone (smog) forming potential* of the speciated hydrocarbon exhaust emissions from biodiesel is 50% lower.
- The exhaust emissions of *carbon monoxide* (a poisonous gas and a contributing factor in the localized formation of smog and ozone) from biodiesel are 50% lower.
- The exhaust emissions of *particulate matter* (recognized as a contributing factor in respiratory disease) from biodiesel are 30% lower.

• The exhaust emissions of *sulfur oxides and sulfates* (major components of acid rain) from biodiesel are completely eliminated.

• The exhaust emissions of *hydrocarbons* (a contributing factor in the localized formation of smog and ozone) are 95% lower.

• The exhaust emissions of *aromatic compounds* known as PAH and NPAH compounds (suspected of causing cancer) are substantially reduced for biodiesel compared to petroleum diesel. Most PAH compounds were reduced by 75% to 85%. All NPAH compounds were reduced by at least 90%." (National Biodiesel Board 1998)

Others have compared the health effects of biodiesel and petroleum diesel as well. One 3.5-year study was sponsored by the US Department of Agriculture (USDA) and the US Department of Energy (DOE). It focused on the life cycle costs and benefits of biodiesel and petroleum diesel. Some of its findings included:

• Overall life cycle emissions of carbon dioxide, a major greenhouse gas, from biodiesel are 78% lower than the overall carbon dioxide emissions from petroleum diesel.

• Emissions of particulate matter, recognized as a contributing factor in respiratory disease and a human health hazard, were 32% lower than emissions from petroleum diesel.

• Overall life cycle emissions of carbon monoxide from biodiesel were 35% lower than overall carbon monoxide emissions from petroleum diesel. When used in buses, biodiesel reduced carbon monoxide emissions by 46%.

• The amount of particulate matter soot in bus tailpipe exhaust was reduced by 83.6% by using a B20 biodiesel blend.

• The use of B20 results in the elimination of life-cycle emissions of sulfur oxides (a major component of acid rain) from bus tailpipes. The DOE study noted, "Biodiesel can eliminate SOx [sulfur oxides] emissions because it is sulfur free" (Sheehan et al. 1998).

• Methane, one of the most potent greenhouse gases, is reduced by 3% when compared to the overall life-cycle methane emissions from petroleum diesel. Though these reductions may appear to be small, they can be significant when estimated on the basis of CO_2-equivalent warming potential.

• The overall bus tailpipe emissions of hydrocarbons, which contribute to the localized formation of smog and ozone, are 37% lower for biodiesel than for petroleum diesel. This property of biodiesel allows for a beneficial effect on urban air pollution.

• The production of biodiesel also results in 79% less wastewater than does the production of petroleum diesel.

• Over its life cycle, biodiesel results in 96% less hazardous solid waste than does petroleum diesel over its entire life cycle. (USDA and US DOE 1998)

When sourced, produced, and used properly, biodiesel is an overwhelmingly cleaner alternative to petroleum diesel. The figures below illustrate the life cycle emissions for B100 and B20 as compared to petroleum diesel and the life cycle carbon dioxide emissions of biodiesel as compared to petroleum diesel.

There is some debate as to whether or not burning biodiesel results in higher emissions of nitrogen oxides (NOx) as compared against petroleum diesel emissions. While one study noted that using B100 in urban buses raised the life cycle NOx emissions, some prominent researchers in the biodiesel field have stated that

NOx emissions do not increase with the use of biodiesel. In addition, technologies are being developed to reduce NOx emissions. These include retarded fuel injection timing, current NOx control technologies (e.g., catalytic converter), and fuel additives (e.g., cetane enhancers and antioxidants) (McCormick et al. 2003). Using hybrid-electric technologies also reduces NOx emissions.

In 2003, the National Renewable Energy Laboratory (NREL) published a report on emissions resulting from a heavy-duty petroleum diesel fleet that used a 20% biodiesel blend. Based on its finding that the maximum rise in ozone was .26 ppb (well below 1 ppb), NREL determined that using biodiesel would have no measurable, negative affect on 1-hour or 8-hour ozone attainment (Morris et al. 2003). In 2006, NREL published the results of another study, which indicated that vehicles using B20 do not have higher NOx emissions than those using petroleum diesel (NREL 2006).

AVERAGE BIODIESEL EMISSIONS COMPARED TO CONVENTIONAL DIESEL, ACCORDING TO EPA		
Emission Type	B100	B20
Regulated		
Total Unburned Hydrocarbons	-67%	-20%
Carbon Monoxide	-48%	-12%
Particulate Matter	-47%	-12%
Nox	10%	2%
Non-Regulated		
Sulfates	100%	-20% *
PAH (Polycyclic Aromatic Hydrocarbons) **	-80%	-13%
nPAH (nitrated PAH's) **	-90%	-50% ***
Ozone potential of speciated HC	-50%	-10%

*   Estimated from B100 result
**   Average reduction across all compounds measured
*** 2-nitroflourine results were within test method variability

Average Life Cycle Air Emissions for B100 and B20 Compared to Petroleum Diesel, According to the US EPA. *Adapted from USDA and US DOE 1998.*

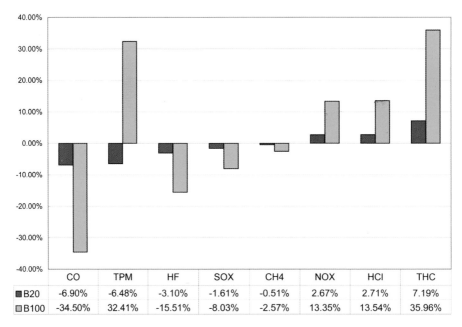

	CO	TPM	HF	SOX	CH4	NOX	HCl	THC
■B20	-6.90%	-6.48%	-3.10%	-1.61%	-0.51%	2.67%	2.71%	7.19%
☐B100	-34.50%	32.41%	-15.51%	-8.03%	-2.57%	13.35%	13.54%	35.96%

Differences in emissions for B100 (100% biodiesel) and B20 (20%/80% biodiesel/petroleum diesel mix) vs. petroleum diesel. *Adapted from the National Biodiesel Board 2007h.*

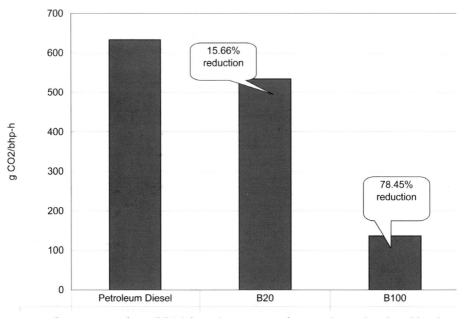

Comparison of net CO2 life cycle emissions for petroleum diesel and biodiesel blends (g CO2/bhp-h). *Adapted from USDA and US DOE 1998.*

Biodiesel Can Be a Domestic Energy Source

Biodiesel can help to reduce our dependence on imported petroleum. Such dependence is both an economic and a strategic problem for the United States and other nations for several reasons, including: the amount of remaining petroleum is limited, US petroleum use is rising, US petroleum production is dropping, US imports of petroleum are rising to make up the difference between production and consumption, and the US transportation sector depends almost entirely on petroleum.

Despite climate change, high fuel prices, and problems associated with energy dependence, the United States' fossil fuel use is on the rise. In 1977, the US imported nearly half of the oil it required to power its industries, homes, vehicles, and so forth. Since that time, it has spent billions of dollars looking for new energy sources and working toward energy efficiency in multiple sectors. Yet, just twenty years later, it broke its own record for dependence on foreign oil imports (Kendell 1998).

In 1999, the Department of Commerce began to investigate the national security implications of this high dependence on oil imports. The resulting report stated that the US already has developed most of its easily accessible, lower cost oil deposits (Bureau of Export Administration 1999). Furthermore, it stated that oil production in the US has decreased steadily and that the country requires an increasing supply of oil imports to meet growing consumption (US Department of Commerce, Bureau of Export Administration 1999). Other oil deposits, such as any found in the Arctic National Wildlife Refuge (ANWR) and certain portions of the Outer Continental Shelf, would provide only a minimal amount of fuel.

Because the US imports most of its oil, low oil prices benefit the economy as they keep costs lower, which often leads to more consumer spending, which leads to a greater gross domestic product (GDP). As oil prices continue to rise, the US economy experiences a slowdown in growth. Because the transportation sector is 97% dependent on petroleum and is the fastest growing energy sector, using a fuel that can be sourced, produced, and used domestically – such as biodiesel – can help to stabilize our nation's energy supply (Greene 1996).

Biodiesel Is Efficient

Producing biodiesel is more energy efficient than producing petroleum diesel. A joint USDA and US DOE (1998) study calculated the amount of embodied energy in biodiesel – in other words, the amount of energy required to grow and harvest the fuel crop, or feedstock, and produce and use the biodiesel. It found that the life cycle fossil energy ratio for commercially produced soy biodiesel is 3.21:1 while the ratio for petroleum diesel is .83:1. That means that for every unit of energy needed to produce biodiesel, 3.21 units of fuel energy results. Meanwhile, petroleum diesel results in a net loss of energy throughout its life cycle; it takes more energy to make it than it yields. Be aware that the embodied energy of biodiesel depends heavily upon the way in which the biodiesel is produced. For instance, biodiesel produced domestically using mustard seed is likely to yield more energy – and be more sustainable – than a biodiesel produced from imported palm oil. We will discuss the importance of sustainable feedstocks in the next chapter.

The figures below illustrate the fossil energy requirements associated with the biodiesel and petroleum diesel life cycles.

Different oil-bearing plants require different amounts of energy to produce. They also yield varying amounts of energy. For example, when compared with petroleum diesel, soybean oil scores well with regard to its energy-efficiency ratio. In contrast to soy-based biodiesel's 3.21:1 energy yield to energy input ratio, petroleum diesel requires 1.1 units of fossil resources to make 1 unit of petroleum diesel (Sheehan et al. 1998). So, in addition to all of its other drawbacks, petroleum diesel is a net energy loser. Biodiesel's energy-efficiency ratio also surpasses that of some other potentially renewable alternatives, such as hydrogen.

Biodiesel produced from waste cooking oil has an even greater fossil energy balance than biodiesel made using virgin oils. That is because the waste cooking oil often is collected from local restaurants, and the resulting biodiesel is used in vehicles and equipment located near the production facility. These factors decrease the energy cost of transporting the oil and fuel to and from the biodiesel plant. For instance, the biodiesel produced by Piedmont Biofuels, which is discussed later, has been estimated to have an energy-efficiency ratio of 7.8:1. (For additional information, see http://biofuels. coop/education/energy-balance/.) That means that for every unit of fuel used for production, biodiesel production yielded 7.8 units of fuel. Producers can maximize their energy balances by insulating reaction equipment, using efficient pumps, and incorporating solar/ renewable fuel heating into the process.

Stage	Fossil Energy (MJ per MJ of Fuel)	Percent
Domestic Crude Production	0.572809	47.75%
Foreign Crude Oil Production	0.539784	45.00%
Domestic Crude Transport	0.003235	0.27%
Foreign Crude Transport	0.013021	1.09%
Crude Oil Refining	0.064499	5.38%
Diesel Fuel Transport	0.006174	0.51%
Total	1.199522	100.00%

Fossil energy requirements for the petroleum diesel life cycle. *Adapted from USDA and US DOE 1998.*

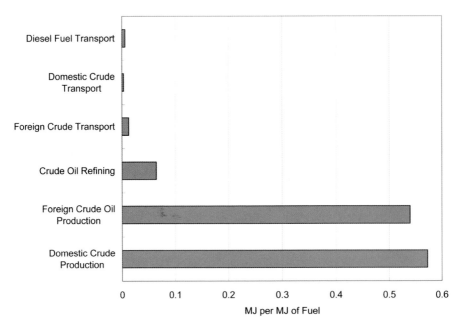

Ranking of fossil energy demand for stages of the petroleum diesel life cycle. *Adapted from USDA and US DOE 1998.*

Stage	Fossil Energy (MJ per MJ of Fuel)	Percent
Soybean Agriculture	0.0656	21.08%
Soybean Transport	0.0034	1.09%
Soybean Crushing	0.0796	25.61%
Soy Oil Transport	0.0072	2.31%
Soy Oil Conversion	0.1508	48.49%
Biodiesel Transport	0.0044	1.41%
Total	0.3110	100.00%

Fossil Energy Requirements for the Biodiesel Life Cycle. *Adapted from USDA and US DOE 1998.*

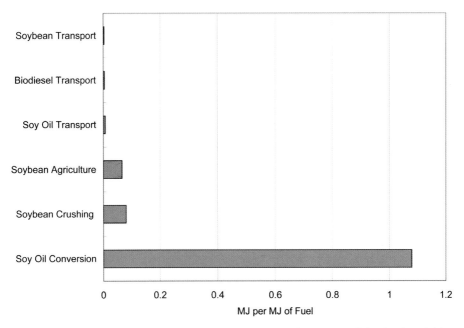

Ranking of primary energy demand for stages of the biodiesel life cycle. *Adapted from USDA and US DOE 1998.*

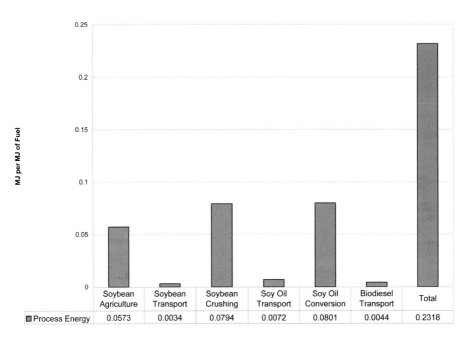

	Soybean Agriculture	Soybean Transport	Soybean Crushing	Soy Oil Transport	Soy Oil Conversion	Biodiesel Transport	Total
▣ Process Energy	0.0573	0.0034	0.0794	0.0072	0.0801	0.0044	0.2318

Process energy requirements for biodiesel life cycle. *Adapted from USDA and US DOE 1998.*

Biodiesel Easily Replaces Petroleum Diesel

Any petroleum diesel vehicle will run on biodiesel. Because it can be operated in any petroleum diesel engine with little or no modification to the engine or the fuel system, biodiesel is an easy replacement to petroleum diesel. There are a few things to keep in mind if you are thinking of running biodiesel in your vehicle:

• Biodiesel that will be used as fuel must meet ASTM D 6751. If you purchase biodiesel for your vehicle, be sure to run only ASTM specification biodiesel in your engine. Those who make and use their own fuel in small-scale plants take personal responsibility for the longevity of their own equipment. Non-certified fuel generally will void engine warranties. It is unlawful to sell fuel labeled as "biodiesel" without ASTM certification.

• Biodiesel not only burns cleaner but also is a great solvent. Biodiesel's cleaning ability can help to keep the fuel line and engine clean. However, if you drive a car that has used petroleum diesel, biodiesel's solvent ability might release deposits that have accumulated in the fuel system during previous petroleum diesel fuel use. The release of these deposits can clog fuel filters. Therefore, check fuel filters often – particularly during the first several months of using biodiesel – and replace them as needed. It is advisable to carry a spare fuel filter in the biodiesel-powered vehicle and to have the tools and knowledge necessary to change the filters on the road. Once the biodiesel has released the fossil fuel deposits from your engine and fuel system, the fuel filters should not need to be replaced as often.

• Because biodiesel is such a great solvent, it can cause any rubber hoses and gaskets in your fuel line to corrode. This problem often occurs in older (pre-1994) vehicles that used rubber rather than synthetic parts – particularly with B100 fuel. Your mechanic or vehicle manufacturer should be able to tell you if you have rubber hoses or seals that should be replaced prior to using fuel blends with higher amounts of biodiesel (e.g., B50, B100).

• Biodiesel is NOT the same thing as straight vegetable oil (SVO), which can be used as a fuel in petroleum diesel engines but only under certain conditions and using special equipment like Elsbett conversion kits. SVO does not meet the ASTM D 6751 specifications.

Some commercially available products purport to make an alternative to biodiesel by mixing used cooking oil with gasoline and proprietary additives. This practice might result in engine damage and is not recommended.

Biodiesel Is Engine Friendly

On January 18, 2001, the US Environmental Protection Agency published the final rule on Heavy-Duty Engine and Vehicle Standards and Highway Diesel Fuel Sulfur Control Requirements (US EPA 2001). The rule stated that diesel vehicles that are ready by model year 2007 needed to cut harmful pollution by ninety-five percent. In order to allow for modern pollution-control technology to be effective on trucks, buses, and other vehicles, the sulfur content of petroleum diesel fuel must be lowered. The Agency required a 97% reduction in the sulfur content of highway petroleum diesel fuel from 500 parts per million to 15 parts per million. The standards were based on the effect that sulfur has on catalytic exhaust emission control devices and other emissions control devices. This dramatic reduction in the sulfur content of petroleum diesel reduced the fuel's natural lubricity, translating into excessive engine wear.

Because biodiesel contains virtually no sulfur, it is a perfect fit for the new low-sulfur regulations, and it has excellent lubricating properties. Without modification, soy-based B100 meets the US EPA's sulfur requirements. Additionally, a biodiesel blend that contains as little as 1-2% biodiesel can increase lubricity by up to 65% (National Biodiesel Board 2001). Although the biodiesel industry hopes to play a significant role in helping refiners meet specifications for ultra-low sulfur petroleum diesel, new additives have been developed to increase lubricity and inhibit corrosion. These additives are being used at the refiner level prior to retail sales of ultra-low sulfur diesel (ULSD). However, some fleet managers have reported lubricity issues and are blending ULSD with low level blends of biodiesel to ensure appropriate lubricity.

Not only is biodiesel an effective lubricant, but it also is a superior solvent and has good burn characteristics. Its solvency helps keep the engine clean, thereby reducing maintenance costs. Meanwhile, it burns well due to its high cetane and oxygen ratings.

Biodiesel Can Be Carbon Neutral

Biodiesel has a closed carbon cycle. In nature, plants photosynthesize, meaning that they absorb carbon dioxide and emit oxygen in order to produce energy for themselves. During metabolism, plants use that energy, thereby using oxygen and releasing some of the carbon dioxide that they stored. Thus, there is a general balance between the cyclical absorption and release of carbon dioxide and oxygen.

Burning biodiesel for fuel emits about the same amount of carbon dioxide (CO_2) that is emitted by burning petroleum diesel. Yet, because biodiesel is made of plants that sequester carbon dioxide during their lives, biodiesel combustion produces *no net increase* of CO_2 in the atmosphere. (Biodiesel made using animal fat has a different carbon dioxide profile, which depends on the type of fat in question, how the animal was raised, what it was fed, and so forth.)

This is unlike petroleum diesel, which is obtained by moving fossil fuels from the lithosphere – beneath Earth's crust – into the biosphere. As a result, when petroleum diesel is burned, it actually adds to the amount of CO_2 in the atmosphere. When people burn fossil fuels, this carbon cycle is thrown out of balance. That is because fossil fuels were made from ancient plant or animal deposits that were buried and compressed over time until they formed coal, oil, or natural gas. Located in the lithosphere (beneath Earth's crust), these ancient deposits only contribute to the atmospheric carbon cycle when they are mined (brought into the biosphere) and burned.

The release of carbon dioxide (CO_2) into the atmosphere from human activities, such as burning fossil fuels, contributes to global climate change. Carbon dioxide levels have risen significantly in our atmosphere since humans began burning fossil fuels for energy. In fact, since the beginning of the industrial revolution, carbon dioxide concentrations have risen considerably. The Intergovernmental Panel on Climate Change (IPCC) has reported that atmospheric carbon dioxide concentrations have increased from approximately 100 parts per million (ppm) in pre-industrial times to 369 ppm in 2005 (IPCC 2007).

In 1999, the US transportation sector became the country's largest source of carbon dioxide emissions, trumping the industrial sector for the first time (US Department of Energy, Energy Information Administration 2004b). Between 1990 and 2002, CO_2 and greenhouse gas emissions per unit of gross domestic product (GDP) dropped by nearly 19% and 21.3%, respectively (US Department of Energy, Energy Information Administration 2004b). This is largely due to improved vehicle technologies that resulted in cleaner-burning vehicles. Offsetting this, in part, is the increased number of miles driven, which has resulted in a net increase of CO_2 emissions. This becomes even more apparent when energy-related CO_2 emissions are represented graphically as shown in the figure below.

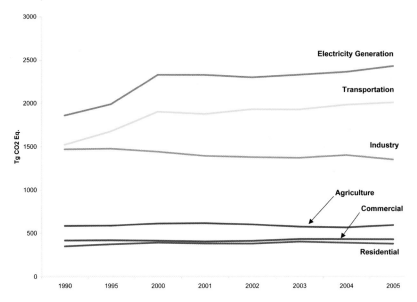

Greenhouse gas emissions allocated to economic sectors in teragrams. *Adapted from US EPA, http://www.epa.gov/climatechange/emissions/downloads06/07Trends.pdf.*

Biodiesel emissions result in a drop in carbon dioxide and methane. Just as we need oxygen to live and grow, plants "inhale" and hold CO_2. When oil from the plants' seeds is converted to biodiesel and burned, that same amount of CO_2 is released into the atmosphere where it can be absorbed by other plants. Thus, biodiesel combustion does not raise the overall amount of CO_2 in the air.

This cannot be said of fossil fuels, which also contain carbon but, because they are stored underground, actually add to the overall amount of CO_2 in the atmosphere when they are taken out of the ground, processed, and burned.

However, because energy from fossil fuels generally is used to produce biodiesel, the recycling of CO_2 with biodiesel is not 100%. Substituting biodiesel, solar, or another renewable energy source for petroleum diesel reduces biodiesel life-cycle CO_2 emissions by 78%. A 20% biodiesel/80% petroleum blend reduces CO_2 by almost 16%. The percentages may seem low, but they pack

The neutral CO_2 cycle of biodiesel. *Courtesy of Briggs et al. n.d.*

quite a punch. In contrast, petroleum diesel requires 100% new CO_2 to make and emits another 100% more when it is used. As a result, the petroleum diesel life cycle emits 178% more CO_2 than does the life cycle associated with biodiesel.

Biodiesel Provides a Way to Recycle Waste Oil

Biodiesel can be made using recycled waste oils and greases. These can be the byproducts of slaughterhouses and food preparation activities and can include rendered fat, tallow, used cooking oil, trap grease, and waste water float grease. Grease in the food industry comes from butter, lard, vegetable fats and oils, fryers, meats, nuts, and cereals and is commonly a regulated waste (Environmental Protection Agency (Queensland) 2006). These feedstocks might have a high free fatty acid (FFA) content and create a challenge when using traditional methods for biodiesel production. However, a number of options exist for converting these feedstocks to biodiesel (Canakci and Van Gerpen 2001).

The lowest valued waste is the one that causes the biggest environmental challenge. Waste fats, oils, and grease (FOG) are the result of hot oil poured down a drain or greasy wastewater entering the drain from a dishwasher. One gallon of FOG can contaminate one million gallons of water (Green Oasis Environmental Incorporation 2002). Grease and oils are fats that are among the more stable of the organic compounds, so they are not easily decomposed by bacteria (NC Division of Pollution Prevention and Environmental Assistance 1999). As a result, cooled FOG congeals with other float greases and coats the inside of drain pipes, pumps, and equipment. Over time, these deposits accumulate and increase in size as more grease and other solid materials build up. The enlarged deposits cause bad odors, slow drainage, overflows, blockages, and equipment failure (see the figures following page).

Cleaning grease deposits from sewers is difficult, expensive, and, sometimes, dangerous (Environmental Protection Department 2006). Often, significant taxpayer dollars go toward cleaning out these sewage facilities, which is a costly municipal problem. It is for these reasons that limits are set by wastewater treatment authorities on how much FOG can be contained in food service wastewater. To help alleviate the problem, building codes usually require the installation of grease traps in restaurants (Environmental Protection Department 2007).

Grease all the way up to the manhole outside a restaurant.
Courtesy of the City of Longmont, CO 2007.

Waste grease buildup being pumped out of a sewer. Courtesy of Philadelphia Fry-o-Diesel, Inc. http://www.fryodiesel.com/trap_grease.htm.

Grease traps are devices that are designed to collect oil and grease before it enters the wastewater stream and to allow for the FOG to be pumped out and collected. Grease traps remove 25-85% (55% average) of grease from wastewater (Peter Crawford n.d.). Usually, the traps are installed underground and placed on kitchen floors or under sinks. Besides biodiesel production, there is little use for trap grease; it is simply a waste product that typically is disposed of at sewage treatment facilities at the cost of five to ten cents per gallon. Although many biodiesel producers use waste oil as their primary feedstock, one company – Philadelphia Fry-o-Diesel – uses trap grease.

Biodiesel could stimulate the installation of superior traps and more frequent pumping because of the potential value added to the waste grease. New grease trap designs separate the FOG and dispense it into a plastic container for removal and emptying into a collection barrel for a rendering company to remove. If such grease traps were installed, a significant amount of waste grease could be collected and turned into biodiesel.

Industrial rendering facilities collect slaughterhouse waste, restaurant grease, and butcher shop trimmings as raw materials. The waste can be rendered for use as biodiesel feedstock or, more commonly, animal feed, composting and vermiculture, soaps (tallow), lanolin, candles, cosmetics, and skin care products (NC Division of Pollution Prevention and Environmental Assistance 1999).

Case Study: Philadelphia Fry-O-Diesel, Philadelphia, Pennsylvania

Philadelphia Fry-O-Diesel (PFoD) is a C-Corporation that was founded in 2004. Currently, it is owned by The Energy Cooperative Association of Pennsylvania, which serves southeastern Pennsylvania with renewable electricity and fuel. After distributing home heating oil for more than two decades and renewable electricity since the deregulation of Pennsylvania's electric industry, The Energy Cooperative wanted to marry its two businesses by finding a renewable heating oil. It found that oil in biodiesel.

But when The Energy Cooperative began looking for a source of biodiesel for its members, it found only rural production facilities centered in the midwestern United States. Believing in "local production for local use," the cooperative wanted to find a renewable energy that could be sourced, produced, and used locally. This would ensure maximum environmental and economic efficiency because source materials and the finished product would not have to be transported long distances. And PFoD was born.

PFoD originally considered using fryer grease for biodiesel production. However, fryer grease is not a waste product, per se, since it has a market and is used as an ingredient for several products, including animal feed (the EU has banned using fryer grease as an additive to animal feed). With a value of about $1.35 per gallon

PFoD decided that, in order to keep the cost of the fuel low, it needed to find a lower value feedstock.

And that it did. It found its source material in the city's abundance of trap grease. For those unfamiliar with trap grease, it is the stuff that fills commercial kitchens' grease traps and sewer lines. By municipal law, food service facilities and restaurants are required to install "grease traps" to minimize the amount of fats, oils, and greases that enter the sewer system. Grease in the sewers is the number one cause of sewer back-ups worldwide. As grease from washing pots and pans and dishes goes down the drain, the grease is "trapped," cooled, and collected. These traps need to be pumped out at regular intervals. Cleaning them out is, well, a not-too-pleasant job as it requires working with a slurry of rotting food, vegetable and/or animal grease, and water.

But besides being really nasty stuff, trap grease also differs chemically from straight vegetable oil and grease used to fry foods. Where straight vegetable oils and even fryer grease and animal fats are composed primarily of triglycerides, trap grease primarily comprises free-fatty-acids. Conventional biodiesel chemistry just won't work.

In 2002, the National Renewable Energy Lab had recognized the need to find new feedstocks for biodiesel production and had identified "brown" greases, including trap grease, as a promising feedstock. NREL estimates that 495 million gallons of trap grease are produced in the U.S. alone every year. According to PFoD, one gallon of cleaned grease will yield one gallon of biodiesel. But NREL had been unsuccessful in getting a pilot plant built to demonstrate the feasibility of producing high quality fuel from this junk.

PFoD rose to the challenge of trying to commercialize this local, problematic waste material. Nadia Adawi, PfoD's President, says, "Grease in the sewers is a universal problem. If you can get grease out of the sewer and make fuel in an urban environment, you've got a win-win on your hands."

PFoD produces biodiesel from trap grease using its own patent-pending technology. With some help from a $370,000 Energy Harvest Grant from Pennsylvania's Department of Environmental Protection (PA DEP), PFoD built a 25,000-gallon/year pilot plant in north Philadelphia. This plant has been operational since late 2006 and is able to consistently produce small volumes of ASTM-spec biodiesel. PFoD has since been awarded a second PA DEP grant in the amount of $250,000 to support development of a commercial production facility.

Nadia says, "Pennsylvania has been incredibly good" in assisting PFoD both with financial support and in streamlining regulations for biofuels producers. However, she points to a number of regulatory hurdles that burden small biofuels producers. For instance, PFoD had to register with the EPA as an ultra-low-sulfur diesel refinery. "We're zoned the same as the Sunoco refinery down the street. We've got no air emissions, no water emissions. But we fall into that same categorization. When we went for our air permit, we were supposed to report our air emissions in tons. We reported ours in grams."

PFoD believes that fuel quality is paramount and has a rigorous testing schedule. In-house tests on basic quality parameters are performed for every batch of fuel, and a composite of five batches is sent to an independent laboratory for the full ASTM test panel. PFoD's fuel was engine-tested in June 2007, when over fifty diverse diesel vehicles were fueled with a B20 blend. But PFoD's current plant is intended for feasibility demonstration, not production – PFoD expects to complete a commercial production facility in 2008 and will seek BQ9000 quality certification at that time.

The company is working to become more sustainable. Nadia says, "We're attempting to move toward sustainability in the following ways: by using a waste product as a raw material, by producing fuel close to where it's used so you're minimizing transportation costs, by creating urban jobs. Our process is energy-intensive; there's no way around it. We wrestle every day with how much solar thermal heating we can do, how much we can burn our own fuel in the process to reduce our dependence on fossil fuels. We use electricity and heat and steam, and we are actively looking for ways to generate what we need with the smallest possible environmental footprint."

Philadelphia Fry-o-Diesel has received licensing requests for its technology from around the world and is working to develop a licensing strategy. It also is looking for investors who can help to fund this new grease-to-biodiesel technology.

Biodiesel Is Relatively Inexpensive

Biodiesel is a relatively inexpensive fuel when compared with petroleum diesel. Although they might cost about the same at the pump, petroleum diesel has many costs associated with it that are not reflected in the purchase price. For example, the cost of transporting the oil from the area of origination to the United States generally is not reflected in the pump price – nor is the energy required to transport the oil to the point of use or the energy lost as a result of an inefficient energy infrastructure. Petroleum diesel's pump price also does not reflect the great expenses incurred to ensure a steady supply of oil from some volatile countries; this can include money spent on defense.

Moreover, the federal government has provided $420 billion in subsidies to the fossil fuel industry (and only a fraction of that in energy efficiency/renewable energy industries) over the past fifty years (Management Information Services 1998). The US General Accounting Office reported that the United States spent more than $130 billion during the last three decades in petroleum industry subsidies. This includes expenditures of $12.5 billion per month on the trade deficit in order to obtain petroleum (Kaptur 2005).

Furthermore, the environmental harm that results from burning petroleum are externalized, meaning that they are not factored into the price of oil (Greene 1996). If the petroleum industry was not heavily subsidized and allowed to externalize the costs of associated harm to the environment (including human health), it would be a very costly fuel. The US large oil imports contribute to our large trade deficit, ultimately reducing the value of the US dollar.

When produced more sustainably, biodiesel does not have the same environmental costs associated with it and does not require the expenditures that are associated with petroleum diesel.

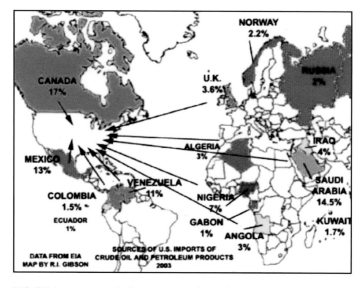

US Oil imports, including imported crude and petroleum products. Based on data from Energy Information Administration for 2003. *Courtesy of US Department of Energy, Energy Information Administration 2003a.*

Biodiesel Stimulates the Domestic Economy

Biodiesel can help to boost domestic and local economies. Using domestic feedstock for biodiesel production is economically advantageous on many levels. On the whole, it is less expensive than importing foreign fuels, and it enables the creation of new jobs. In addition, both food and fuel are made from the biodiesel process when, for example, soybeans are used. By increasing feedstock production, not only does the fuel supply rise, but the supply of meal that is used primarily for animal feed rises as well. Of course, this could lead to a glut, which could have negative repercussions.

The federal and state governments are assisting the biodiesel industry and, thereby, the domestic economy. For instance, Minnesota's mandate that, by 2005, all petroleum diesel fuels sold statewide must contain a minimum of two percent biodiesel has provided a real boon to the state's farmers. On the federal level, the renewable fuel standard has also provided an incentive for the growth of domestic biofuels economies. Authorized under the Energy Policy Act of 2005, the renewable fuel standard requires that a minimum of 7.5 billion gallons of renewably based alternative fuels be mixed with motor vehicle fuel sold in the US by 2012.

It has been estimated that 300,000 new American jobs could be created by the renewable energy industry. A US Department of Agriculture study indicated that 85% of the revenue generated by a renewable energy production facility is spent within a seventy-five-mile radius of the plant and that, for every dollar in revenue generated, a $2.25 overall economic impact is achieved. For states that have no oil/gasoline production, the economic benefits are significantly greater (Bryan 2002). The regional model for biodiesel will help to ensure that biodiesel is not only an "alternative" to fossil fuels but that it is environmentally preferable.

Case Study: Ian Heatwole, Farmer, Weyer's Cave, Virginia

The Heatwole family has a long history of farming in Virginia's Shenandoah Valley. This region produces a modest share of soybeans, which ranks eighth in terms of cash production in Virginia for agricultural commodities. In 2005, the state produced 15 million bushels. Ian currently owns and operates Fox Run Farms where he produces 8,000 bushels of soybeans annually. In addition, he maintains 500 heads of dairy cattle, 11,000 heads of turkeys, and 48,000 broilers.

The idea of producing biodiesel came to him about seven years ago when rumors of "soydiesel" came to the area. At the time, he was trying to find a way to get energy from all of the dairy and poultry manure produced on his farm. He says, "I have always tried to maximize the use of products I produce on my farm, so to minimize things I need to buy and use inputs as efficiently as possible." Ian began looking into anaerobic digestion as a way to utilize the manure. He did plenty of research, but a broken leg hampered his ability to tinker with machines and systems.

By the time he was well enough to get into the shop, spring had come, and there was little time to build. He had an acquaintance who was very pro-biodiesel at the time. They would get into "debates" about methane digestion vs. biodiesel. Meanwhile, diesel prices had reached $1.50/gallon. Ian uses a diesel engine to power an irrigation pump. After a brutal summer during which he ran the pump for nearly 2000 hours (at over 5 gallons of diesel per hour), Ian started doing his research into biodiesel. Says Ian, "Look at my farm. I grow soybeans. I use diesel. I use soybean meal. If someone can buy my soybeans to make soybean meal and oil, and someone can buy soybean oil to make biodiesel, and everyone is making (slim) profits along the way, then I should be able to do all that myself cost effectively."

In fact, farmers in the Shenandoah Valley must ship their soybeans 200 miles to the nearest soybean crushing facility, which is located in the eastern part of the state. That is where most of the state's soybeans are produced so that is where the plants are located. This fact largely explains why Virginia has essentially excluded the Valley from biodiesel considerations. Although the state has put incentives in place to facilitate the development of the biodiesel industry, the Shenandoah Valley has no chance to meet the 10 million gallon per year minimum requirement set forth by the state incentives (this threshold was recently reduced to 2 million gallons). The result is major energy expenditures for farmers to transport their crops out of the valley. In addition, the soybeans that are taken to eastern Virginia are processed into high-protein animal feed, much of which is sent back to the Shenandoah Valley where the majority of Virginia's dairy industry is located (another 200-mile trip). The turning point for Ian was when he realized he could use every product and byproduct of the process to replace inputs that he currently purchases. "Abandon (temporarily) methane digestion, and the biodiesel sickness begins," he exclaims.

Ian is now in the process of developing an on-farm soybean processing/biodiesel production facility. He is using a processor developed by RASP Technologies that has the capability to produce up to 250,000 gallons of fuel annually using locally grown feedstocks. He plans to start out processing 60,000 bushels of Shenandoah Valley-grown soybeans into roughly 75,000 gallons of fuel. In addition to providing fuel, these 60,000 bushels will not have to make the energy-intensive 200 mile trip to eastern Virginia, and the resulting meal will not make the 200 mile return trip back to the Valley. Both the bushels and the meal ordinarily are transported by truck and must cross the Blue Ridge mountain range, which is an energy-intensive process. Factoring in fuel consumption, about 8,000 gallons of diesel fuel per year could be saved from soybean/meal transportation alone. Ian will use the fuel as a B50 blend on the farm and sell the rest to area farmers through a local petroleum distributor. The same holds true for the protein meal byproduct.

Thus, the energy savings incurred by the project are several-fold:

- It reduces the amount of petroleum diesel fuel that will be needed by local area farmers by displacing current consumption with clean-burning, renewable energy.
- It reduces the amount of fuel wasted in the transportation of soybeans to the region's only soybean crushing facility 200 miles away.
- It reduces the amount of fuel wasted on the return trip that brings high-protein meal back to dairy farmers that live in the Shenandoah Valley.

These energy savings are in addition to providing a market for locally grown crops and providing animal feed for local livestock producers.

Ian's model of on-farm biodiesel and animal feed production is one that is being replicated all over the world. With rising fuel costs, more and more farmers are realizing that it is economically favorable to produce their own fuel and animal feed using crops already grown on their farm. They are eliminating the middleman and creating a sustainable system that has long-term viability. Energy costs are one of the single largest factors that influence the price of food. Therefore, when farmers make themselves immune to rising fuel prices, they are keeping prices down for end consumers.

Adds Ian, "Efficiency is important to me, so it's what makes it the number one reason and what keeps me trudging down this path (thankfully, the trudging is almost over). The environmental, economic, and engine life benefits are icing on the cake."

That is some pretty good icing.

Biodiesel Has Many Uses
Heating Oil

Biodiesel can be used as heating oil in home or industrial oil furnaces. Number 2 heating oil is very similar to petroleum diesel vehicle fuel. Biodiesel can be mixed with heating oil in a B20 blend and burned in an oil-fired furnace or boiler; some people have experimented with higher blends as well.

The price for petroleum fuels has been extremely low in the past, leaving little interest in other alternatives. However, higher heating costs in recent years have become a concern for many. Recent developments overseas as well as hostility and volatility in many oil-bearing countries have sensitized informed individuals and companies to consider other options.

"Bioheat" is the term used when biodiesel is used for heating a home, business, or any other space. The benefits of using biodiesel in home heating applications (or in boilers) are many. First of all, home heating oil is usually of poor quality and contains sulfur, which can dirty heating oil burners and boilers, lower the overall efficiency of the heating system, and increase maintenance costs. Lower efficiency means higher fuel consumption and less heat per unit of fuel. Biodiesel is a cleaner fuel that contains virtually no sulfur and, therefore, requires less maintenance. The National Renewable Energy Laboratory sponsored a National Biodiesel Board study in Warwick, Rhode Island, that demonstrated that biodiesel blends of up to 20% have a positive impact on the cleanliness of filters, strainers, and nozzles. This is important because filters, strainers, and nozzles have chronic premature failure (National Biodiesel Board 2007b). Besides sulfur contributing to higher maintenance costs, sulfur in fuel oil also translates to sulfur dioxide (SO_2) emissions. Petroleum diesel fuel and heating oil may contain as much as 5,000 ppm of sulfur. As a result, large SO_2 reductions can be realized by blending biodiesel into fuel oil/off-road petroleum diesel fuel.

Another study indicated that home heating systems burn fuel differently than engines do; as a result, their emissions are different. When used in home heating systems, biodiesel emits a lower amount of NOx than conventional home heating oil. A B20 blend reduces NOx emissions by 20% over the entire range of air settings as compared to conventional home heating oil. Sulfur dioxide emissions are reduced by 83% (Krishna 2003; Batey 2002).

On a life cycle basis, biodiesel fuels emit 78% less carbon dioxide than number 2 fuel oil emits. If a carbon dioxide credit system is implemented that is similar to the SO2 market-based system created by the US Environmental Protection Agency, biodiesel might be used toward those credits.

Bioheat appears to be compatible with all known oil tanks and systems at blends of 20% or less. The compatibility of higher blends will depend on the material used in your own personal tank, pumps, and fuel lines. For blends higher than 20% biodiesel, the only recommended vessels are those made of aluminum, steel, stainless steel, mild steel, fiberglass, fluorinated polyethylene, or fluorinated polypropylene.

Some materials, such as copper, tin, zinc (galvanized), lead, brass, or bronze, are not recommended for tanks or liners because they can clog filters or cause sediment to form. Fluorinated polyethylene, fluorinated polypropylene, Teflon, Teflon-lined, or Viton components are recommended for gaskets, seals, hoses, and O-rings in blends with more than 20% biodiesel (National Biodiesel Board 2007a). It is recommended that a) low level blends of biodiesel are introduced and b) a step-up program is used for a bioheat transition because home heating systems that have burned heating oil tend to have significant buildup of sediment and sludge.

Biodiesel demonstrates high solvent properties, providing a cleaning effect that tends to dissolve or loosen some sediment that has been deposited in tanks and fuel systems from years of home heating oil use. Since biodiesel tends to dissolve the sludge that often coats the insides of old fuel tanks and fuel lines, a clogged fuel filter or burner head might occur if the first few trials are not monitored closely. Biodiesel can be added as a lower percentage blend until sludge in the fuel tank dissolves. It is worth keeping an extra fuel filter on hand during the first heating season.

The best way to store biodiesel for home heating use is in an indoor or underground storage tank because biodiesel, like number 2 heating oil, will gel if stored outside in extremely cold weather. The pour point for number 2 fuel oil is -24° C (-11° F). The pour point for biodiesel varies somewhat depending on feedstock and concentration but is consistently much higher than that of number 2 fuel oil. Temperatures above this pour point would be necessary for storage. The USDA Agricultural Experiment Station estimated that if everyone in the northeastern United States used a blend as low as 5% biodiesel, 50 million gallons of regular heating oil per year would be saved (Pahl 2005). Greg Pahl's book *Biodiesel: Growing a New Energy Economy* provides a strong overview of biodiesel as a home heating oil. The National Biodiesel Board's web site also speaks to biodiesel as a home heating oil option as well.

Biodiesel also has demonstrated success as a kerosene substitute. Kerosene, also called diesel number 1 or number 1 heating oil, is thinner than biodiesel. Blending biodiesel at lower percentages with kerosene usually works well.

Power

Temporary backup diesel-fueled generators often are used in emergency situations; however, they lack an after-treatment process for reducing the harmful emissions in their exhaust. As a result, backup generators located indoors can pose a particularly strong health threat. The toxicity associated with petroleum diesel compounded with a confined space and little ventilation can be hazardous to humans and others. Using B100 in a generator will eliminate nearly all hydrocarbons and sulfur emissions. Carbon monoxide emissions are reduced by about half, and particulate matter is reduced by over two-thirds (NBB 2007j). (However, we do no advocate running a B100-powered generator indoors.)

Another potential market for biodiesel electrical generation is in remote areas where communities use diesel generators as their sole source of electricity. Most of these isolated communities have to import all of their energy fuel sources, which can be very expensive when you consider the current price of petroleum. Thus, rural communities may benefit from an electricity-generating fuel that can be grown, produced, and used locally.

California's Sierra Railroad Company and others are taking advantage of biodiesel's versatility and environmental benefits. The company has started a project that it calls PowerTrainUSA, which will allow cities, municipal utility districts, or industrial power users to avoid costly blackouts and to lighten the electrical load for the state through distributed generation. Under a five-year peaker power contract, 48 General Electric locomotives will be used for 1,000 hours annually and will use about 7.5 million gallons of biodiesel annually (Hart n.d.). (A peaker power plant is one that runs only when there is high demand for electricity.) Sierra Railroad Company's President Mike Hart has stated that if 48 parked locomotive engine electrical generators were connected to the state's power grid and run on biodiesel, there would be enough power during peak demand hours to support roughly 100,000 homes (Jensen 2002).

Cleaning

Since biodiesel is a solvent, it can be used as a paint stripper, a means to clean the sludge out of tanks that have been used for number 2 diesel fuel, or as a general cleaning agent but without the offensive smell that petroleum diesel has. However, due to its natural solvent capability, biodiesel should be removed from surfaces when cleaning is finished or when spilled on painted items accidentally.

Marine Applications

Biodiesel can be used effectively in marine applications as well. It is non-toxic, free of sulfur, aromatics, and the additives that petroleum diesel contains. It also biodegrades readily and, after 28 days in an aqueous solution, can degrade by 95 percent (versus 40 percent for petroleum diesel fuel). Biodegradability is an important factor when considering the damage that can occur after an oil spill.

Biodiesel made from virgin vegetable oils has been tested as a cleaning agent for shorelines contaminated by crude oil spills. Scientists at the School of Ocean Science at the University of Wales tested and approved biodiesel for this application. In fact, vegetable oil methyl esters similar to biodiesel have been licensed by the California Department of Fish & Game as a "shoreline cleaning agent" to extract crude oil from shorelines and marshes after a spill. This oil spill cleanup biosolvent, CytoSol, was developed by CytoCulture (Wedel 1999). When an oil spill does occur, vegetable oils or biodiesel can be used to clean birds and other animals without stripping them of their body oils (ENVIRON Corporation 1993).

Furthermore, many diesel boat operators have reported a noticeable change in exhaust odor – for the better – when they have used biodiesel, and biodiesel does not result in the same eye irritation that petroleum diesel exhaust can cause.

Biodiesel can be used in a variety of applications in the marine industry, including recreational boats, inland commercial and oceangoing commercial ships, research vessels, and the US Coast Guard fleet. Individuals and organizations throughout the world are taking note of these facts and utilizing this renewable, domestic fuel in their marine fleets.

All Biodiesel is Not Created Equal

Several years ago, biodiesel began gaining a reputation as an inexpensive, do-it-yourself fuel that could be made from used cooking oil. Small-scale producers began scouring their neighborhoods for restaurants from which they could collect used oil. The idea of inexpensive diesel fuel intrigued some farmers, who essentially could grow their own fuel, planting and processing feedstocks like soy or canola and converting that virgin oil into biodiesel.

Fast forward to today. The biofuels revolution is underway, and, with it, growing concerns over the environmental impacts of some biofuel. Biodiesel has transitioned quickly from being a small-scale, do-it-yourself project that can be made and used locally to a global money-maker. Some governments and companies see the biofuels revolution as a means of capitalizing on the world's energy demands.

What does this mean? It means that biodiesel can be divided into two basic categories – 1) locally or regionally sourced, produced, and used and 2) globally sourced, produced, and used. These two categories are quite different with regard to their environmental impacts. For instance, it takes far more energy to transport the biodiesel feedstock to the production plant and distribute the finished biodiesel than it does if your feedstock and customers are local or regional. Using waste vegetable oil instead of virgin feedstock requires even less of an energy input and does not require land to be cleared for farmland or farmland to be used for fuel crops rather than food crops.

Thus, all biodiesel is not created equal. And this is where a tension is emerging. For some, biodiesel represents a viable alternative that can help to support local economies while providing environmentally cleaner fuel for existing diesel vehicles. For others, biodiesel represents the new gold rush. Poorer countries with few natural resources suddenly see economic opportunity in the biofuels economy.

For instance, some small farmers around the world see biodiesel crops as more lucrative than food crops and are glad to be able to support their families by using agricultural land to grow biodiesel fuel crops, or feedstocks. Meanwhile, some large companies and governments are seeing green – and not the environmental kind – in the large-scale production of biodiesel crops. Particularly for poorer countries with little or no petroleum deposits, the idea of participating in and becoming wealthy from the production of crops for biodiesel production is too tempting to refuse. Thus, the economic opportunities that are associated with biodiesel and other biofuels (e.g., ethanol) have provided some growers with a short-term financial incentive to destroy habitat and replace it with single-crop plantations of biofuel feedstocks, such as palm trees for palm oil. Rainforests and other ecosystems are being destroyed and replaced with plantations of single crops that in no way offer the ecological complexity and benefits afforded by natural ecosystems.

The rainforest's plants play a vital role in storing carbon dioxide. Destroying rainforests and other ecosystems for plantations of biofuel crops results in the release of carbon dioxide into the atmosphere, thereby exacerbating – rather than improving – the human impact on global climate change. So, if you are buying biodiesel because it has lower carbon dioxide emissions than petroleum diesel fuel, beware that biodiesel made from rainforest-produced feedstock can do more harm than good. As climate scientists have suggested, "Biofuels are likely to speed up global warming as they are encouraging farmers to burn tropical forests that have absorbed a large portion of greenhouse gases" (Nakanishi 2007). Scientist James Lovelock, known for the Gaia theory, stated, "Some of these alternative energy schemes, such as biofuels, are truly dangerous. If exploited on a large scale, they will hasten our downfall" (Nakanishi 2007).

This raises an important issue regarding renewable fuels. That is that *renewable fuels are not necessarily environmentally friendly*. The environmental preferability of biodiesel – or any other alternative fuel – cannot be based on emissions alone; it must be based on the environmental impacts of the fuel *over its entire life cycle*. Clearly, not all biodiesel is created equal.

The Purpose of This Book

We wrote this book for two reasons. First, we wanted to consolidate and share the current biodiesel best practices, safety information, quality standards, regulations, and incentives. There are many producers out there – particularly small-scale producers with limited resources – who would like to be able to find all of this information in one place. We hope that this book satisfies that need.

Second, we wanted to help move sustainability from the edges to the center of the biodiesel dialogue. Despite some gains in biodiesel industry, from the human and environmental health perspective, many of the best practices, specifications, regulations, and incentives are too piecemeal and narrowly focused to ensure the sustainability of biodiesel.

Therefore, this book is both a resource guide that is intended to provide information on human and environmental health and safety information, laws, and incentives *and* a book that explains that they do not go far enough to result in sustainable biodiesel.

We discuss the need for more sustainable feedstocks and explore the best practices related to biodiesel safety, storage, and waste disposal – all from the perspective of human and environmental health. Then, we discuss the push for quality biodiesel with the ASTM specification, which is required for commercial biodiesel, and with the development of BQ9000. We highlight the environmental regulations and permitting that relate to biodiesel. Then, we explore the need for sustainable biodiesel certification and guidelines and throughout this book, we provide several case studies of those working in the biodiesel industry.

Chapter 2. Biodiesel Feedstocks

This chapter discusses biodiesel feedstocks. First, it introduces the many plants that can be used as feedstock and indicates the amount of vegetable oil and animal fats that are produced annually in the United States. Next, it discusses the growing demand for biodiesel feedstocks and some of the human and environmental devastation that is occurring around the world as a result.

Feedstock Basics

Biodiesel can be made from any number of oil-containing materials, both plant and animal. Any plant that yields oil can be used as a feedstock for biodiesel production. These include everything from algae to macadamia nuts to sunflowers to rapeseed (canola) to soybeans. Animal fats leftover from rendering plants and fats (animal and vegetable) caught in restaurant grease traps also can be used to make biodiesel. Most small-scale biodiesel producers make fuel from used cooking oil recovered from restaurants. The table below lists some of the many plants that can be used to make biodiesel.

Table 1. Some Biodiesel Feedstock Materials

algae	coriander	linseed (flax)	pecan nuts
avocado	corn (maize)	lupine	poppy
brazil nuts	cotton	macadamia nuts	pumpkin seeds
calendula	euphorbia	mustard seed	rapeseed/canola
cashew nut	hazelnuts	oats	safflower
castor beans	hemp	olives	sesame
cocoa (cacao)	jatropha	opium	soybean
coconut	jojoba oil	palm	sunflowers
coffee	kenaf	peanuts	tung oil

In the United States, soybean oil is the most common commercial biodiesel feedstock while oil palms are grown for use in the tropics, and rapeseed (canola) is used in the cooler parts of Europe and Canada.

Biodiesel, alone, cannot replace all of the petroleum diesel fuel used in the United States; however, it can serve as part of the nation's alternative energy portfolio. In fact, Charles Peterson, reporting on a study by of the University of Idaho, stated, "It would be very ambitious" to replace just the amount of petroleum diesel used on farms in the U.S., which is about 3.1 billion gallons. Furthermore, this would require a predicted 15% of our total production land area. Peterson further contends that it would even "be very ambitious to have a 0.5 billion gallon per year biodiesel industry," which would only replace 1.5% of our on-highway petroleum diesel use and even less of our fuel oil and kerosene consumption.

Table 2. Total Annual Production of US Fats and Oils

Vegetable Oils	(Billion pounds/yr)
Soybean	18.340
Peanuts	0.220
Sunflower	1.000
Cottonseed	1.010
Corn	2.420
Others	0.669
Total Vegetable Oil	**23.659**

Animal Fats	(Billion pounds/yr)
Inedible Tallow	3.859
Lard & Grease	1.306
Yellow Grease	2.633
Poultry Fat	2.215
Edible Tallow	1.625
Total Animal Fat	11.638

(Source: Pearl 2002)

Nonetheless, those who are interested in global ecological sustainability might have concerns over the use of non-sustainable (e.g., non-organic, monocrop, till-intensive) farming methods to create a "greener" fuel source. Therefore, the focus should be on ensuring that the entire life cycle of biodiesel is environmentally sound while, of course, conserving energy. Certainly, if biodiesel was made from oil crops grown organically, it would provide an even more favorable energy balance throughout the biodiesel life cycle. The need for a Sustainable Biodiesel Label is being addressed by the Sustainable Biodiesel Alliance.

Crops that are grown and used to make biodiesel also can be used as food crops. The American Soybean Association reports that one bushel of soybeans comprises about 18.5% oil, 35% protein, and 5% fiber. After these soybeans are processed, they provide about 11 pounds of oil and 48 pounds of protein meal, which can be used for human/animal consumption (ASA 2005).

Terrestrial crops are not the only potential source of oil for biodiesel production. Algae is another. A program funded by the US Department of Energy's Office of Fuels Development indicated that algae can be grown using waste carbon dioxide emitted by coal-fired power plants. Additional research is taking place around the world on the potential of using algae is a biodiesel feedstock.

Case Study: Yokayo Biofuels, Ukiah, California

Yokayo Biofuels is an "S" corporation that began as a partnership in October 2001. Initially, the founders considered starting as a cooperative but decided against it. As founder Kumar Plocher states, "Not everyone is cut out for do-it-yourself projects like making fuel." In his experience, cooperatives might have many enthusiastic members who do not necessarily have the right skill set. He suggested that a biodiesel cooperative has to be very well run in order to be successful. Instead, Kumar found that having a corporation helps him to deal with the regulatory and other issues that his company faces.

Initially, Yokayo Biofuels intended to market and distribute biodiesel but not to produce it. In the fall of 2001, Kumar found a way to get a load of biodiesel from Connecticut-based World Energy to his area in northern California. Since that time, the company has grown to ten employees and forty-five shareholders. And it produces biodiesel using waste oil collected from about 500 area restaurants.

In California, the collection of used fryer oil is regulated by the California Department of Food and Agriculture (CDFA). At regular intervals, CDFA sends a Meat and Poultry Inspector out to Yokayo Biofuels' production plant to make sure that it continues to operate within the confines of its Inedible Kitchen Grease Transport License as well as its Rendering License (which allows the company to process the oil). A big hurdle for most people who would want to collect oil from restaurants legitimately is the $2 million liability insurance requirement (in addition to vehicle insurance). Other than that, Kumar says, it is pretty basic. Surprisingly, it is a felony to collect grease without a license.

Yokayo's business model relies on almost a closed loop biodiesel life cycle since the company controls collection, production, and distribution. The company makes about 300,000 gallons of biodiesel annually, which it sells to its 100-200 customers. Kumar says that the company delivers the biodiesel itself using a 2,000 gallon delivery tanker and that it will add a 1,500 gallon delivery truck soon. It also sells through several retail pumps, although most of its business is not pump-based. Households also are encouraged to bring their waste cooking oil to Yokayo for use in biodiesel production.

Kumar states that the company has done well but faces challenges in the area of financing. It has shied away from venture capitalists because, he says, they tend to infiltrate the business plan and want to see unending growth. However, this does not reflect Yokaya Biofuel's vision and business model. The company's vision is to establish a successful business model for decentralized, local fuel production and distribution. Kumar does not plan to continuously grow the company in a centralized fashion. The company is likely to max out production at one to three million gallons per year. Unless another sustainable oil source comes along, it is not likely to be able to produce much more than that. Once it has achieved success, the company can franchise this model out and copy it in other places or let others do it.

Kumar states that there is no reason that a locality should ever neglect to make full use of its regional resources. Unfortunately, that is exactly what is happening all over our society. An example of this can be seen in agricultural subsidies that encourage mono-cropping. Another example is found in petroleum subsidies that encourage an addiction to oil that gets shipped from around the world. His vision is pretty simple: find what the community can sustainably supply as a feedstock (source oil) for biodiesel and, then, work with it to create a local economy that fuels itself – independent of outside influences. Ideally, every ingredient and input for the fuel is produced locally, including labor and investment dollars. The money generated by this local fuel economy is kept in the community, so the community thrives. Additionally, there is no longer a need to worry about rising oil prices affecting the life and livelihood of the community's inhabitants. People become empowered.

Because biodiesel can be made from an almost infinite number of feedstocks, it stands to reason that any region should have an ideal feedstock – all we need to do is find out what it is. In California, the ideal "starter" feedstock is clearly used fryer oil, or "UFO" as Kumar's father calls it. It is plentiful and typically is processed with other byproducts to make low-grade animal feed. A substitute for petroleum diesel is clearly a much more value-added application. However, because restaurants have been conditioned to pay renderers for the service of used oil removal, they are more than happy to have their oil taken for free by a biodiesel company while at the same time gaining the public relations benefits of doing something politically correct and environmentally correct.

This situation will not last. Eventually, all of the used fryer oil will be accounted for by biodiesel companies, either directly or indirectly, and restaurants will be charging for its pickup. At that point, a more suitable local feedstock will need to be found if the local fuel economy is to grow further. In Yokayo's case, the company is committing research and development resources to various oil "super crops," including Chinese Tallow Tree, Jatropha, and various microalgae species. The goal is to find something that will grow well in the company's area and yield at least ten times the paltry amount of oil that you get from an acre of soybeans, which is 50 gallons per acre. At 500 gallons an acre, it begins to look possible to supply all of a community's diesel needs with biodiesel. Kumar says that the Chinese Tallow Tree generally is regarded as yielding between 500-1000 gallons per acre, Jatropha oftentimes is placed anywhere between 200 and 1000 gallons per acre, and microalgaes can be all over the map, with some varieties yielding about 900 gallons per acre and other yielding as much as 10,000 gallons per acre.

Yokayo Biofuels does not run into many regulatory challenges at the state and federal levels but finds itself working to comply with county issues at its new site. Being a small company, Kumar says that he is working with the county regulators, who have both created some regulatory hurdles for the company and been very supportive.

At a roundtable meeting with the local fire district, Environmental Health, and Building and Planning regulators, Kumar asked for more information on a particular topic. The fire chief threw his arms in the air and remarked, "We've never had a biodiesel producer in this area before." It was a way of saying, "I don't know." Kumar says that his company frequently hears that from local regulators. On the one hand, the regulators know that Yokayo Biofuels is a chemical plant because making biodiesel is a chemical mixing process. On the other hand, the regulators also can see that the company is much smaller and, in many ways, more benign than just about any type of chemical plant in operation. Yokayo Biofuels uses only three ingredients (mostly vegetable oil, with smaller quantities of methanol and potassium hydroxide), and its product is non-hazardous and non-toxic. Everything is done at low temperature and pressure. It is all much simpler than a refinery or glue factory or water treatment plant or whatever it is regulators are used to seeing.

However, Kumar understands the value of regulation. He says that all of the regulators can sense that there are both dangerous and benign aspects to their business, but they do not always know where to look or what it is they are seeing. From his perspective, Kumar wants to ensure that his company ends up with a production plant that he can show off, eventually with tours open to the public. Such a scenario requires a set of best practices, and that is what the company is working to create with the local regulatory authorities. There is a lot of trust involved. He trusts that they will not view him as a miniature petroleum refinery and try to shut him down, and they trust that he will give them honest information about what exactly it is that Yokayo Biofuels does.

The company is strict about the quality of its product, testing every batch of biodiesel for compliance with the ASTM standard and sending samples weekly to a third party for testing. It is a big proponent of fuel quality testing and wishes that there was federal enforcement of biodiesel quality so that the auto manufacturers would feel comfortable with biodiesel use.

Safety also is a concern, and Yokayo has developed policies that others – particularly home brewers and small-scale producers – might face. For example, after witnessing firsthand fires created through the spontaneous combustion of oil-soaked rags, the company uses many special containers for rags and other oil-soaked items.

Yokayo Biofuels does not receive much financial support. In order to qualify for the federal tax credit, its sells B99.9 biodiesel. This tax incentive provides 50 cents per gallon sold. "That's the one bit of financial help that we get," Kumar says. Yet, biodiesel producers using waste oil are at a disadvantage because producers who use virgin soy oil get 99.9 cents per gallon of fuel sold despite the potential environmental preferability of waste oil-based biodiesel. Kumar attributes this to the power of the soy lobby.

Finally, Kumar says, "I see a lot of starry-eyed people who think that they can save the world and become rich at the same time. I've been doing this for six years, and I'm making the same $15/hour that my employees are making." He adds, "If you're doing it right, if you're doing it according to ideals, you're up against a lot.... If you're serious and you're in this for idealistic reasons, there's not much help."

Growing Demand for Feedstocks

As the global demand for biodiesel grows, large areas of habitat are being cleared and replaced with biofuel crops. This is occurring in several regions, such as Latin America and Asia. For instance, farmers in Colombia are being driven from their land by paramilitary groups who view biofuel crops as less risky than growing coca, from which cocaine is derived. Some Colombian companies are creating false deeds to lay claim to farmland throughout the country, thereby creating one of the worst refugee crises outside of Africa (Blach and Carroll 2007).

There is a similar situation in Paraguay where farmers are being forced off their land for soy (soya) plantations. While soy is used for food, the current push for the production of soy oil for biofuels is driving up the demand for farmland on which soy can be planted.

To meet its biodiesel demand, Argentina plans to expand its soybean production, which will have serious environmental and social repercussions. First, soybean farming is replacing far more biologically diverse, ecologically productive habitat. For instance, the Pizarro nature reserve in Argentina's northwestern Salta province lost its status as a reserve when the regional government decided to auction off part of it to agribusiness companies (Valente 2007b). It took nearly two years of campaigning before the land sale was cancelled and the reserve regained its legally protected status. Although Pizarro was spared, other villages have not been, and the soybean jobs that have moved into new areas often are given to outsiders, not locals. Second, to meet the demand for biodiesel, farmers anticipate having to expand the soybean production area by about 10%. Yet, some of the land that has not previously been used to grow soybeans is not as fertile as farmland already under production, which means that it will require more land to grow the same amount of soybeans. Third, some of the most productive soybean farming regions are those in which extreme poverty has grown the most. Fourth, when monoculture biofuel crops replace several existing crops, this can hurt family farms and force seasonal workers and smaller farmers into cities to look for work. Land ownership, meanwhile, is concentrated in the hands of large companies who control the biofuel feedstock.

About 28 million tons of palm oil are produced worldwide every year, making up more than one-fourth of the world's vegetable oil production (Roundtable on Sustainable Palm Oil 2007). Papua, New Guinea, the world's third largest exporter of palm oil, has experienced social problems resulting from the rise in plantations. Many of the small-scale plantations that once were managed by indigenous or village landowners who did not hold formal land titles have been taken over by large corporate plantations. Working conditions on these large-scale plantations are not good (Dietzen

2007). This has eroded the self-sufficiency once experienced by the small-scale oil producers.

Meanwhile, in Southeast Asia, palm oil plantations are replacing rainforest. This has resulted not only in the release of greenhouse gases (e.g., carbon dioxide) into the atmosphere, but it also has resulted in forest fires and the sedimentation of water bodies. For example, Indonesia's rainforests contain more than half of the world's tropical peat. Peat is rich in carbon and methane, both greenhouse gases, which are released into the atmosphere when burned. In the race to plant biodiesel oilcrops, Indonesia's rainforests are being burned down in their conversion to farmland. Largely because of this, Indonesia is the third largest carbon dioxide emitter in the world. In fact, one study reported that for every ton of peatland-grown, palm oil biodiesel, between ten and thirty tons of carbon dioxide are emitted (Climate Ark 2007). In July 2007, Wilmar, the world's largest palm oil distributor, came under scrutiny from Friends of the Earth International (FOEI). FOEI stated that Wilmar is responsible for destroying Indonesia's rainforest to grow palm oils for the food, chemical, and biofuels industries. FOEI contended that Wilmar illegally cut and burned down Indonesia's forests (Navarro July 2007a).

Some energy-intensive countries are depending on the agricultural abilities of their trading partners to supply them with enough feedstock to meet their biofuels goals. For instance, The Renewable Energy Road Map released by the Commission of the European Communities suggests that Europe's demand for biofuels can be supported, in part, by Europe's trading partners – particularly those in developing countries. Though such trade could be a boon to the economies of developing countries around the world, some concerned with sustainable biodiesel have suggested that biodiesel feedstocks should be used first for the countries producing them; only if there is excess should it be exported for trade.

The Need for Sustainable Feedstocks

Even in those places in which land is not stolen or deforested illegally, the replacement of habitat with monocrops is troubling for two reasons: 1) most farms are not as ecologically beneficial as natural habitat (for information on the problems inherent in most agriculture, read publications by The Land Institute at www.land-institute.org) and 2) forests and other habitats play a key role in sequestering carbon dioxide, a greenhouse gas, and keeping it out of the atmosphere. When forests or other natural habitats are razed and carbon dioxide is released into the atmosphere as a result, the human impact on global climate change is worsened rather than improved. Biodiesel produced from feedstocks grown under these conditions is more an environmental burden than a benefit.

Therefore, despite the wide range of feedstocks available for biodiesel production, the biodiesel industry is facing an enormous challenge – the lack of *sustainably grown or collected* feedstocks. This issue is highlighted by the previous discussion in which the production of biodiesel feedstocks is creating rather than resolving environmental problems, such as climate change. Yet, the environmental degradation associated with feedstocks is not just an international problem; it also is a domestic one. Thus, sustainability needs to be central to the discussion of both domestically and internationally produced feedstocks.

Most agriculture is non-sustainable for many reasons, including the initial conversion of complex habitat to simple farmland; the tilling of soils, which breaks up the communities of microflora and microfauna; the frequent use of chemical pesticides and fertilizers on non-organic farms; and the development and use on hybridized and genetically modified seeds.

Can biodiesel be sustainable when some of it is made from non-sustainably produced feedstocks? If we are truly committed to sustainability throughout the entire biodiesel life cycle, the answer is "No." Thus, if the biodiesel industry decides that it wants to be sustainable as a whole, it will need to work to green its supply chain. Part of greening the supply chain means changing the way in which we do agriculture.

Conservation Agriculture: A Way Forward

When geneticist Wes Jackson co-founded The Land Institute in 1977, he had an unusual notion about farming. Jackson believed that the till-based agriculture that began about 10,000 years ago and that is practiced around the world today is environmentally non-sustainable.

From the ancient Greek and Mesopotamian civilizations to modern agricultural societies, Jackson has said that humans have practiced the wrong kind of farming. Most farming is based on the idea of planting annuals, harvesting them seasonally, and, then, tilling the soil and replanting. When plants are taken out of the ground, the "services" that their roots provided – such as taking in and storing water and providing habitat for insects and important soil microorganisms – end abruptly. Tilling causes soil communities to be broken up, and the soil loses precious water and nutrients and becomes vulnerable to erosion.

In his article "Natural Systems Agriculture: A Radical Alternative," Jackson explains that archaeological evidence has shown that farming-related soil erosion and deforestation have been problems in Greece during the past 8,000 years. Both Plato and Aristotle noted the ecological degradation that resulted from agriculture. In his dialogues, Plato described the loss of Greece's rich, deep soils and the land's resulting inability to hold once-abundant rainwater: "Once the land was enriched by yearly rains, which were not lost, as they are now, by flowing from the bare land into the sea." Sumer, Babylon, Assyria, and Rome fared little better. On the other side of the world, in the highlands of central Mexico, much of the soil was lost 3,500 years before the 1518 arrival of Cortez.

Nearly three decades ago, Jackson set out "to develop an agriculture that will save soil from being lost or poisoned while promoting a community life at once prosperous and enduring." The Land Institute calls its type of farming Natural Systems Agriculture. On 370 acres of farmland in the middle of the U.S., Jackson has sought to create a method of farming modeled after the native prairie ecosystem – one that enhances biological diversity while satisfying human needs.

It is a way of farming based on the idea of perennial polyculture, which is how nature farms. The Institute harvests foods without digging up their parent plants and breaking up soils. It grows both domesticated wild perennials, which are naturally adapted to local conditions, and perennial versions of major grain crops. It also plants crops in polycultures rather than monocultures; this mixing of crops helps to prevent losses from agricultural pests.

So far, The Land Institute's results have been promising. Plant yields are about equal to those of conventional farming without the associated environmental damage. And it seems as though Jackson's ideas may be catching on as others experiment with his farming method in different ecosystems. "We don't know how this is all going to turn out," he stated in an interview with writer Scott Russell Sanders. "But the risky thing is to do nothing, to keep going the way we've been going."

Chapter 3. Biodiesel Safety, Storage, and Waste Disposal

Because biodiesel is intended to be an environmentally preferable type of fuel, all aspects of its production and use should have a minimal environmental footprint. However, if safe production practices and proper waste disposal are ignored, this environmentally friendly fuel can be harmful to both environmental and human health. This chapter explores the key safety issues, including fire safety, human health, storage regulation, and waste disposal.

Human Health and Fire Safety

Human health and safety is a vital factor when producing biodiesel on any scale. Each production facility, regardless of size, must take certain steps to ensure the safety of all those working around and within the facility. Chemical poisoning, fire, burns, explosions, and other accidents are very real dangers, which can result in serious injury, death, and loss of property.

Lyle Estill, founder of Piedmont Biofuels in Pittsboro, North Carolina, states that safety has nothing to do with size or whether a biodiesel producer has enough money to take appropriate safety precautions. "It has to do with the people involved," he told us. "I have seen dangerous, deeply unsafe, large biodiesel, and I have seen deeply unsafe small biodiesel. There are some producers who are safety conscious, and they pay attention and they care." Making biodiesel does carry risks, and, without taking the appropriate precautions, it can be dangerous.

"We've had some people die – one in South Carolina," says Lyle. "Home brewers have burned their backyard sheds and their houses to the ground. And we've had large commercial operations burn to the ground." In fact, one large plant in California burned down because methanol burned in non-groundable plastic totes.

These and other issues highlight the importance of safety – both for people and the environment – in biodiesel production.

Methanol is a key ingredient in biodiesel production – it is the alcohol that often is used to make biodiesel while lye is the catalyst. Ethanol can be used as the alcohol in place of methanol but generally is not due to its higher cost, feasibility in the reaction, special challenges, and a potentially unfavorable energy balance. The flash point for methanol is 54°F while the flash point for ethanol is 62°F, making ethanol slightly less flammable than methanol. Additionally, working with ethanol is less dangerous with regard to human health. The National Fire Protection Association (NFPA) health rating for ethanol is 0 (the lowest) while methanol is rated as 1. However, one should apply the same safety practices to ethanol in order to err on the side of caution.

There are many safety issues that must be considered when working with methanol. The key concerns of methanol are its flammability and potential effects on human health. Many federal and local guidelines have been established to minimize the risk of fire and to reduce exposure to humans. These guidelines are set forth by the NFPA. Additionally, some organizations, such as Methanex, provide information about the safe use and handling of methanol. These guidelines vary from the storage containers to protective gear worn by employees who work around the methanol. Specific locales might have their own regulations regarding fire safety, so it is worthwhile to check with the local fire marshal to see if your area has any special rules and regulations. Chapter 5 discusses this further.

Because of methanol's flammability, fire safety in a biodiesel production facility is particularly important. Methanol has a NFPA fire rating of 3, which means that it is ignitable under almost all ambient conditions. Fires can be prevented by keeping the methanol away from any source of open flames, sparks, and oxidants. Excessive heat also can cause methanol to ignite so it should be stored in a cool place. If methanol vapors leak from tanks and accumulate in a confined space, they can explode. Methanol vapors are heavier than air and may settle in low points. If vapors collect near pilot lights or sparking motors, they could be ignited inadvertently. Tanks containing methanol also can explode if they are exposed to high heat for an extended period of time.

Fighting methanol-based fires can be difficult due to the fact that methanol flames are nearly invisible in direct sunlight. The only way to detect the flames in this situation is to search for heat sources as well as any other material that is visibly burning.

Methanol fires can be extinguished with oxygen deprivation measures, such as dry chemical powder, CO_2, and alcohol-resistant foam. In the event of a fire, methanol containers must be kept cool by spraying water on them. Firefighters attacking the blaze need to wear breathing devices and clothing that will not allow any vapors to reach their bodies (Methanol Institute 2006). Buildings, cabinets, and containers that store methanol must be labeled "Flammable Liquids" to alert emergency personnel in case of an accident.

Methanol is harmful if it is ingested, inhaled, or absorbed through the skin. The general rules are as follows: do not ingest methanol (even small amounts can be fatal), avoid skin contact, and avoid prolonged exposure to vapors. Of course, you should seek medical help immediately if any of these situations occurs. Because methanol vapors can be harmful, the exposure limits for working with methanol are eight hours if the concentration is below 200 parts per million and fifteen minutes if the concentration is above 250 parts per million (Methanex n.d.).

Generally speaking, if a person can smell methanol, that person is over the safe exposure limit. According to the US EPA, "Methanol has a slightly alcoholic odor when pure and a repulsive, pungent odor when in its crude form; it is difficult to smell methanol in the air at less than 2,000 parts per million (ppm)" (US EPA 2000b). Any activity that results in methanol vapors entering the workspace should be eliminated from the processing routine. Production equipment should be of a closed-system design so that the transfer and handling of methanol and methanol containing liquids (biodiesel, glycerol, and sometimes wash water) occurs in a fumeless manner. When fluids are transferred from tank to tank, air evacuated from the tanks should be vented to the outdoors rather than into the workspace.

Some of the symptoms of methanol overexposure are headache, weakness, drowsiness, nausea, difficulty breathing, eye irritation, blurred vision, drunkenness, unconsciousness, and even death. It only takes four ounces or less of ingested methanol to cause permanent nervous system damage, blindness, or death. Other effects of ingestion include poisoning, systemic acidosis, damage to the optic nerve, damage to the central nervous system, and degreasing the skin, which leads to dermatitis.

There must be proper ventilation in the facility in order to protect from overexposure to this dangerous chemical. The type of ventilation system will ultimately depend on the characteristics of the facility. For example, Gaston County Schools in North Carolina produce about 150 gallons of biodiesel per day in a facility that is located in the county's old garage. To get the facility approved, the fire marshal and engineers inspected it. The new system is fully enclosed, is ventilated regularly, and employees are encouraged to wear respirators. (NOTE: Methanol vapors will not be stopped by conventional masks or cartridge respirators. The only safe respirator for methanol vapors is a supplied-air respirator, such as a NIOSH or SCBA respirator. Although most respirators will *not* work for methanol fumes, we do recommend them for KOH and NaOH dust.).

To prevent accidents, the Methanol Institute suggests that anyone working around methanol should be properly trained in how to handle it and should wear protective clothing, such as safety goggles, elbow-length gloves, long pants, closed toed shoes, and a protective apron. To prepare for and react to a methanol spill, develop and implement a spill response plan. In many cases, a methanol spill should result in contacting hazardous material response personnel. Review the materials safety data sheets (MSDS) for further guidelines on methanol safety and environmental response. The same concerns that apply to methanol also apply to ethanol.

Other methanol issues include:
- Bonding and grounding for spark-free transfer is a best practice.
- Store in approved fire cabinet or storage tank.
- Methanol is a regulated air and water pollutant. Minimize any release of methanol to the air and to surface or ground waters.

Lye (sodium hydroxide or NaOH) and caustic potash (potassium hydroxide or KOH) are the two catalysts commonly used in the biodiesel reaction. Both catalysts are strong corrosives that can result in severe burns and blindness. The dust of NaOH and KOH can be harmful or fatal if inhaled in excessive amounts. Small stray lye particles will cause skin irritation and damage clothing and electronic equipment.

Proper safety gear for handling NaOH or KOH includes indirect-vented safety goggles suited for preventing chemical splash injury (safety glasses do not suffice); dust mask or cartridge respirator; elbow-length, chemical-resistant gloves; long pants; closed shoes; and a protective apron. Work in a well-ventilated space when measuring catalyst to prevent dust exposure.

Lye absorbs water from the atmosphere, which creates a handling challenge. In the event of a spill, lye can be neutralized using vinegar.

Obtain, read, and post materials safety data sheets for methanol, NaOH or KOH, and any other hazardous chemicals used in the biodiesel production facility (see CU Biodiesel's web site at http://www.cubiodiesel.org/links.php for pertinent MSDS).

Keep all hazardous chemicals securely locked when not in use, and ensure that they are labeled at all times. Untrained personnel, children, and pets must not have access to chemicals, lab ware, or any production components that have contacted lye, methanol, or other compounds. There are other process chemicals and other substances used in biodiesel production as well as risks associated with byproducts containing these chemicals. Below is a table of substances that may be used in your production process and the type of risk associated with each substance.

Table 3. Substances Used in Biodiesel Processing and Their Associated Risks

Item of Risk	Type of Risk
Vegetable Oil/Animal Fat	Flammable, falls due to surface spills
Methanol	Poison, flammable, blindness due to splashing
Sodium Hydroxide	Caustic
Potassium Hydroxide	Caustic
Sodium Methoxide (liquid)	Caustic, flammable
Potassium Methoxide (liquid)	Caustic, flammable, blindness due to splashing
Sodium Methoxide (crytalline)	Caustic, poison, highly flammable, reacts violently with water
Potassium Methoxide (crystalline)	Caustic, poison, flammable, reacts violently with water
Phenolpthalein Reagent	Poison, flammable
Isopropanol Alcohol	Poison, flammable

(Adapted from Kemp 2007, 381)

Case Study: Matt Steiman, Dickinson College, Carlisle, Pennsylvania

Matt Steiman, the biodiesel project supervisor at Dickinson College, is an enthusiastic educator. He began his biodiesel journey when he was an employee of Wilson College seven years ago. Currently, he runs Dickinson College's biodiesel operations, which are part of the College's sustainability initiative. The sustainability initiative focuses on raising organic foods for the college at the Open Sky Farm, operating a dining services composting program, teaching students about sustainable agriculture, moving forward with the American College & University Presidents Climate Commitment, supporting the protection and restoration of watersheds, showcasing Dickinson's LEED-certified Center for Sustainable Living, and many more programs.

The biodiesel program at both Wilson and Dickinson College have been successful due, in part, to Matt Steiman's commitment and enthusiasm. Matt has written and been awarded a few grants, including a $5,000 award from the Community Environmental Legal Defense Fund and the Pennsylvania Farmers Union, to build a biodiesel processor and to use the fuel in Wilson College's Fulton Farm's irrigation pumps. The processor also was used as the catalyst for six workshops held in 2006 for farmers in remote locations around Pennsylvania. The workshops were hands-on, full-day biodiesel experiences. Farmers were introduced to biodiesel and, then, were able to make their own small batches, later working with Matt to make a large batch of fuel. After the workshops, the project continued, and grant money was distributed to ten farms throughout Pennsylvania to assist farmers in building their own biodiesel reactors. Matt served as the technical expert and support person. Farmers were given plans for the "Appleseed" reactor (designed by Maria "Mark" Alovert), which was the reactor design recommended; they also were allowed to design their own reactors and process biodiesel from waste grease that they collected from local restaurants.

Matt also secured two other grants from the Pennsylvania Department of Environmental Protection to teach teachers about biodiesel.

He took about thirty teachers through a biodiesel production lab experiment, which now is used as part of the normal curriculum. The largest grant that Matt helped to secure was from the USDA's Sustainable Agriculture Research and Education (SARE) program and has helped the biodiesel program to grow tremendously. Matt hosts student groups, sets up field trips to his plant, gives guest lectures on biodiesel, assists with environmental science laboratories and wrote a lab exercise, attends fairs and events, and many other activities. Matt states, "My job is to make fuel and teach about it."

Matt says that Dickinson College makes about 50-100 gallons of fuel per week, which the College uses in campus garbage trucks, lawnmowers, tractors, and other equipment. The feedstock is waste oil collected from about eight area restaurants and another local college at a rate of about 50 gallons per week. He says that it is a "pretty laid back" model in which they give the restaurant a barrel and, when the barrel is full, the restaurant calls the College, which picks it up. Dickinson has 1,000 gallons of storage on-site, which has worked well. The biodiesel project at Dickinson is housed in the facilities management building.

Matt was hired by Dickinson College in January 2006, about a year-and-a-half after the College began tinkering with a biodiesel program. It built a small Appleseed reactor but was unable to figure out staffing and launch the program. Since then, the program has taken off and hosts state interns and work study staff. During the summer of 2007, Matt had two full-time students working on research projects and making fuel. The physics and biology major interns helped to develop a solar heating system for their biodiesel plant and make soap from the glycerin sidestream (after the methanol was recovered), which they hope to use in campus bathrooms and give out at events.

Matt considers one of his biggest contributions to the biodiesel movement in Pennsylvania to be his work with the regulatory bodies in the state. He is working with the Department of Environmental Protection (DEP) and health and safety professionals to develop official guidelines on what the clear regulatory and permitting requirements are for small-scale biodiesel producers. Matt says, "As a teacher, it's important to know what is okay, safe, and environmentally responsible." Matt had contacts from the Department of Environmental Protection, as well as other contacts from the environmental field, who have assisted in his endeavor. He invited a representative from the Waste Management Division to visit the plant to learn what biodiesel is all about.

To date, the DEP has required no permits and allows the group to compost the glycerin at Open Sky Farm. Matt hopes to do some research in the future that will assist the DEP in developing regulations for the disposal/reuse of biodiesel byproducts, such as how composted glycerin actually affects a compost pile.

Matt worked with the campus health and safety officer, who gave the biodiesel group fire extinguisher training and made fuel with the group. At the time, they were "taking a break from methanol recovery," but, after the health and safety officer explained that the methanol fumes were above the lower explosion limit for methanol and above the safe human level, they quickly re-incorporated methanol recovery into their biodiesel production process.

Luckily, their shop involved the design of the whole floor as containment, so any spills are contained in the building. They have had few safety issues but did have an oil-soaked rag spontaneously combust (they were able to contain the smoldering rag quickly to prevent major damage), and a few students knocked over a tank of glycerol only to find trouble getting someone to help them clean it up. They realized that they needed to establish a clear emergency response plan to deal with such issues.

Byproducts are a big issue in biodiesel production, and many new to biodiesel production think little about this aspect until they have accumulated a great deal of waste. Matt said that Dickinson's biodiesel group recently started monitoring the oil that they collect from restaurants. Bad oil is a sidestream, and many people do not know what to do with it. Matt has composted a good amount of it in horse manure piles with success.

When asked about quality testing, Matt states that he hopes to coordinate with the chemistry folks on campus for assistance in that area.

Matt gets calls frequently from potential home brewers. When asked what kind of advice he gives them, he first said, "Think hard about time. If you don't have the time to do it, you are not going to save money." Then, he added, "Have open conversations, promote best practices, keep small-scale biodiesel a legal practice in Pennsylvania. If producing biodiesel in Pennsylvania means you have to recover the methanol, then recover it." Regulatory bodies are important. "Wastewater treatment plants have their role, and biodiesel producers shouldn't flush byproduct down the drain and hope it's okay. It takes a lot to run a shop, and there is plenty of danger that should be taken seriously. Safety is more important than making a single gallon of fuel."

Matt said that someone recently had a biodiesel-related barn fire, which just goes to show that "we should all know what's safe and a good practice and especially environmentally responsible. Fifty percent get into biodiesel because they think they can make cheap fuel, and fifty percent to be green. It's important to know what being green means" and if you make biodiesel with an "interest in sustainability, it's worth the time."

Since Matt has been "behind a table holding up jars of biodiesel in front of thousands of people by now," he feels a responsibility and is excited about the prospect of helping to establish clear answers for biodiesel producers in Pennsylvania so that "if they want to protect themselves from legal enforcement and be safe," they can. But, he says, "It's a very slow process."

Production Facility Safety Considerations

When designing a production facility or a space for your biodiesel production system, you must be cognizant of numerous safety, health, and environmental considerations. Some of these are listed below. Note that this list is not exhaustive and that each producer typically faces varying obstacles related to safety.

Safety Measures

Below lists equipment or products that should be in the biodiesel production facility or workspace:

- Eyewash station
- Fire extinguishers
- First aid kit
- NIOSH-approved or appropriate respirator for methoxide (e.g., SCBA) and catalyst dust
- Explosion-proof trash can (for oil-soaked rags)
- Anti-static straps

Below lists safety measures that should be taken during the biodiesel production process:

- Wear approved protective gloves and eye protection
- Use chemical and fireproof gloves, and avoid using latex gloves that could dissolve after coming into contact with certain chemicals.
- Cover exposed skin and consider an apron used in chemistry labs.
- Label all containers
- Buy lots of Sharpies!

• Prepare a safety and action plan (this can be the SPCC plan discussed in Chapter 5), and ensure that everyone working in the space is familiar with it.

The plan should have prevention as the key theme. A good way to develop a plan is identify each risk presented (MSDS are good for this) and determine ways to prevent or lower the risk as well as an action plan. Ensure that all tools and equipment are acquired so that, if the hazard presents itself, you are adequately able to respond.

• Ensure grounding of all exposed metal components. Grounding provides a method to dissipate electrical energy in a fault condition safely.

• Ensure adequate room and tank ventilation, and activate the room ventilation fan when in use.

Production space should be well-insulated, tightly sealed, and well-ventilated. Multiple chemicals used in biodiesel production will ignite as concentrations build above lower explosive limits (LEL). To ensure that this does not happen, the workspace should be sealed but built with an intake air vent in an area that provides maximum cross-ventilation. A high-capacity ventilation fan can be installed with automatic dampers to enhance the cross-ventilation (Kemp 2006, 385). William Kemp recommends that the room be "tightly sealed with standard construction-grade vapor barrier and acoustical gap sealant and is fitted with a ventilation fan that is capable of drawing approximately 500 cubic feet of air from one end of the room to the other per minute of operation" (Kemp 2006, 268). He contends that this high turnover rate of air keeps methanol vapors far enough below the lower explosive limit (LEL) and keeps the air quality from becoming poisonous. He also suggests the following:

• Prior to welding or cutting tanks that have been used for methanol, fill them with water.

- Methanol vapors are ignitable and explosive. Do not use mixing equipment that will ignite methanol vapors.
- Use of mixing equipment, such as an electric paint-mixing paddle that is attached to a variable speed drill, can ignite methanol vapors since they are very flammable. Instead, use explosion-proof or hydraulic motors for agitation, or alternatively, recalculating air pumps (Kemp 2006, 207).
- Always have running water available.
- Provide secondary containment to prepare for spills or leakage from storage tanks.
- Keep MSDS sheets for all chemicals on site.

Storage of Ingredients, Byproducts, & Biodiesel

Depending on the state in which you operate, biodiesel storage tanks might be subject to regulation. Ensure that tanks are compatible with all of the chemicals/materials that are used in your operation. There are process and waste collection tanks as well as tanks for flammables and corrosives. The process tanks need pressure relief and venting safeguards. Stainless steel tanks are the best option for blending tanks. A professional engineer might need to certify tanks used for glycerol storage.

Tanks storing methanol must be enclosed thereby eliminating any methanol entering the atmosphere; additionally, storage tanks must meet Underwriters Laboratories (UL) 142 standards. UL 142 standards set regulations for aboveground storage tanks that contain flammable materials. For example, UL 142 requires tanks to be tested and inspected for leaks before they are shipped to the customer. Local regulations and site requirements dictate what types of labeling are needed on the tanks that contain methanol. Special procedures should be followed whenever methanol is transferred. The material for the tank is another important consideration. Mild steel, a low-carbon material that has the proper strength and does not react with the materials it contains, is a very common and effective choice.

The storage of biodiesel byproducts is another important issue for which some regulations have been developed. For example, Virginia has determined that storage containers must be made of material that is compatible with the glycerin, methanol, and the catalyst. The containers need to be grounded to prevent them from catching fire from sparks and should remain closed to prevent gaseous releases to the air. Additionally, a spill/leak plan should be in place, and any spills or leaks should be addressed immediately. The minimum amount of material that should be recycled each year is seventy-five percent to avoid speculative accumulation regulation (this is discussed in Chapter 5), which means that if you have 100 pounds of glycerin from your reactions, you must recycle at least 75 pounds from that waste within the year.

Waste Disposal

Glycerin

Glycerin is a waste product that is created during biodiesel production. There appear to be few regulations governing the disposal of glycerin. In the State of New Hampshire, glycerol cannot be sent into the septic or sewer systems (NH Dept. of Environmental Services). However, if the glycerin is absent of free liquids, it can be taken to a landfill.

In its biodiesel production, North Carolina's Appalachian State University recovers its methanol from the glycerin through a distillation process. Recovered methanol goes back into the processor to be used for future reactions. Then, the refined glycerin can be composted or used to make soap.

Meanwhile, Rice University and Colorado University are keeping their glycerol and researching different uses for it. Colorado University is keeping its glycerol as well. The University of Connecticut has a service to remove its glycerol for disposal while North Carolina's Gaston County Schools use the glycerol to make a variety of products, such as cat litter, floor absorbers, and hand soap, and are exploring using it to make antifreeze.

In Virginia, the regulation of the glycerol byproduct depends on whether or not the methanol is recovered. There are regulations in both cases, but there are more in the case in which the methanol is not recovered because it is considered hazardous waste rather than a solid waste. If the methanol is not recovered and the glycerin contains between forty and sixty percent methanol, it is considered a hazardous material, and a protocol for disposal must be followed. A sample of glycerin that contains unrecovered methanol will have a flash point of less than 140°F, which is the catalyst for regulated hazardous waste disposal requirements.

If the methanol is recovered from the glycerol byproduct, it is classified as a solid waste rather than a hazardous waste. Virginia allows it to be sent to a refining plant or disposed of in another appropriate manner (e.g., industrial composting).

An alternative to glycerin disposal is to compost it if your state allows this practice. However, be aware that glycerin generates a large amount of heat similar to mulch and, in dry conditions, can ignite a fire. As a precaution, fire breaks should be built around the composting area. Since glycerin has high carbon content, other substances need to be mixed with it in order for best results. Another, albeit less recommended, method for glycerol disposal is to burn it in a boiler. In order to do this in some states, an individual must prove to the state that it can be done safely and with minimal air pollution. Some producers also have investigated dust suppression, erosion control, and anaerobic digestion. Open dumping is almost always prohibited.

The glycerin from multiple home brewers' biodiesel batches can be combined to justify the expense of methanol recovery if the cost is prohibitive. This will prevent the waste from becoming regulated under hazardous waste regulations and will ensure a much lower environmental permitting burden. Plus, it is the right thing to do from an environmental perspective.

Case Study: Dr. Jack Martin, Burlington, North Carolina

Dr. Jack Martin is "primarily a teacher" who has been involved with renewable energy, including biodiesel, for decades. A graduate of North Carolina's Warren Wilson College, he received his teaching certificate. While there, he was influenced by the College's New Alchemy Institute where he built a wind turbine and a geodesic dome. He went on to receive his Master's degree in fish culture and PhD in sustainable resource management. His postdoctoral work was in agriculture and environmental technology.

While he was in the Peace Corps in Nepal, Jack experienced firsthand the 1978-1979 fuel crisis resulting from the earlier embargo. He converted a motorcycle to run on ethanol so that he could get around. In addition to the embargo, he says, "There are places in the world that can't get fuel oil because we outbid them. India doesn't get much fuel. Nepal gets precious little. Eastern Europe is slammed all the time – Russia stopped sending gas in the middle of winter. In Poland and the Czech Republic, people were burning furniture to stay warm."

Once he returned home, he maintained his interest in renewable energy. In 1998, he met Joshua Tickell, who helped to jumpstart the US interest in biodiesel. The following year, Jack taught a course on Sustainable Resource Management at Appalachian State University. Following Joshua's book *From the Fryer to the Fuel Tank*, Jack's class converted a car to run on biodiesel. The following year, they converted a composting tractor. In the ensuing years, Jack was actively involved with Appalachian State University, and several students brought their diesel vehicles so that they could modify them to run not on biodiesel but on waste vegetable oil.

From 2004 to 2005, he taught a renewable energy class course at Piedmont Community College. This was the origin of Piedmont Biofuels. In 2005, "Sustainability Jack" began co-hosting "Home Power Hour," a radio program, with sidekick "Solar Jim."

A few years ago, Jack met Eric Henry of TS Designs, a North Carolina-based t-shirt company known for its environmental concern. TS Designs had just purchased a biodiesel processor, and Jack knew how to make biodiesel. Eric asked Jack to help them with their biodiesel production. They had to rebuild most of the processor to make it work properly, but it did function.

During their biodiesel processing, they used a compost pile – not only to compost the biodiesel waste byproducts but also to heat the biodiesel process. They were able to bury the tanks or the plumbing in the compost, which maintained a temperature between 160-180 degrees Fahrenheit. The compost pile was "typically at 130-160 degrees." However, Jack says, "The microbes love the waste products of glycerol and bump it up to 160-180 degrees."

Today, Jack teaches courses in Sustainable Resource Management and Sustainable Transportation at Appalachian State University. He believes in "food first and biofuels as a byproduct."

Wastewater Treatment

Wastewater treatment is another important issue to consider in biodiesel production. Although some organizations simply pour their wastewater down the drains (which is not recommended), others have come up with ingenious ways to handle wastewater. For example, Appalachian State University (ASU) has constructed a unique system to treat its wastewater after it is used in biodiesel production. The system uses bioremediation, which mimics a series of ponds with a sandy bottom. The system also includes riparian buffers, bacteria, and algae. After the water goes through these mechanisms, it can be filtered and reused for biodiesel production. The system handles about 360 gallons per week. Its cost was relatively low – in part due to the donation of some of the components. However, ASU has indicated that a system like this could be built for as little as $750. Another unique part of the system is the fact that it takes place in a passive solar greenhouse, which ensures that the system maintains a steady temperature that allows it to operate properly. ASU also uses solar energy to keep its oil storage heated and to heat the oil to its reaction temperature.

Case Study: Blue Ridge Biofuels, Asheville, North Carolina

Blue Ridge Biofuels began in 2003 as a cooperative in Asheville, North Carolina. Its core group of dedicated members began producing biodiesel for themselves on a farm located in the same county. Driving to the farm, producing biodiesel, and returning to Asheville with the finished product was a bit challenging due to the distance involved. The co-op wanted to make the fuel closer to home and to grow into a company that could recycle a valuable commodity (waste vegetable oil) into a saleable product.

Because the company was so young, it could apply for business loans, but its founders had to back those loans. Because they had to take on the financial risk personally, they decided to become a for-profit business – a worker-owned limited liability corporation (LLC). There are six full-time employees, four of which are original founding members, and four part-time employees, including some founding members, and every employee owns part of the company.

Now, Blue Ridge Biofuels focuses on creating the biodiesel infrastructure, producing biodiesel, and delivering biodiesel. It has set up fueling stations, including the first publicly accessible B100 pump on the East Coast. Now, it has two B100 pumps (including one in the nearby town of Black Mountain) and has several other filling stations that sell B20.

It provides about 120 restaurants and factories in Buncombe County with yellow grease bins, which it empties using its 1,000-plus gallon tank vacuum truck. It produces 250,000 gallons of biodiesel each year and is expanding to produce more than one million gallons of biodiesel annually. It also sells a B20 blend of home heating oil. Currently, it has to buy some biodiesel from another North Carolina company in order to keep up with the demand. When it finishes its expansion, it should be able to meet the local demand for biodiesel.

As the company expands, it buys one piece of expensive equipment at a time. It has received some cost-share grants, meaning that the company must fund its own projects and be refunded for part of those costs later. Its first fuelling station was partially funded this way. Owner Mac Minaudo says, "We bought a $30,000 centrifuge. We had to buy that $30,000 centrifuge and wait three months to be refunded."

Blue Ridge Biofuels prides itself on its exemplary customer service and the fact that it sells locally produced biodiesel using locally produced waste feedstock. However, it is looking for ways to reduce its costs and its prices to make it more competitive with petroleum diesel. With B20, Blue Ridge Biofuels can be cost-competitive with petroleum diesel; it is harder to do with B100. Fortunately for small-scale producers, the state of North Carolina dropped its excise tax for driving on straight vegetable oil. It also ended the excise tax for home brewers who use (and do not sell) their own fuel. The state also is considering dropping the 30-cent-per-gallon excise tax on fuel that is produced in North Carolina, which would help Blue Ridge Biofuels' B100 to be more cost-competitive. The state's lost excise tax revenue might well be made up by the growth of biodiesel, which is North Carolina's most widely used alternative fuel, and the infrastructure (e.g., construction) that accompanies a growing market.

Of course, Blue Ridge Biofuels ensures that its biodiesel meets the ASTM specification and is working actively to surpass the ASTM specs. That is because there are some fuel-related problems that can arise even from ASTM-spec fuel. Blue Ridge Biofuels has begun a stringent, in-house quality control program under the voluntary BQ9000 certification program, which looks at how the fuel was made, the feedstock used, and how it was put into the truck.

In order to ensure well-blended fuel, Blue Ridge Biofuels mixes its fuel on delivery day. The fuel is further agitated during driving and blended, again, when it is pumped into the storage tanks at the refueling station. Winter still poses a challenge for blending as the cloud point for biodiesel is lower than that of petroleum diesel, and the two need to be mixed together at a warmer temperature. In order to avoid associated blending problems, Blue Ridge Biofuels stores its biodiesel in heated storage tanks and uses an anti-gel for winter use in its B20 fuel.

Mac says, "We have quite a cool customer base. We sell fuel to the Grove Park Inn, Biltmore House, Asheville Airport. Those are all B20. And we actually do sell B5 fuel to the City (of Asheville) through our partnership with a local petroleum company – Biltmore Oil. Biltmore Oil is amazing," he says. It was founded by a philanthropist who has helped Blue Ridge Biofuels by providing a good price on petroleum for the B20 blend and who also supports the local community. Other customers include the Mountain Horticulture Research Center, the Mountain Horticulture Center (a different organization), Warren Wilson College, the University of North Carolina – Asheville, and the North Carolina Arboretum, which runs on B70 year-round.

With regard to regulations, Blue Ridge Biofuels has been well supported. The company worked with the state's Department of

Transportation for tax-related issues. As for safety issues, Mac states, "There was a learning curve that needed to be established with entities like our fire marshals. Since there wasn't anything on the books about biodiesel… they had to come up with interpretations of the code for biodiesel specifically." He explains that they wanted to treat it like diesel fuel but that there is a great difference between the two in that biodiesel is about two times less flammable than is diesel fuel. He continues, "And, then, we've had some difficulties with the state and taxes. They don't have anything on the books about how to tax biodiesel. So, it became tricky when we were selling biodiesel that we make and selling biodiesel that other people make."

Blue Ridge Biofuels is in full compliance regarding safety. Mac says, "We work very closely with our fire marshals as well as the city engineers to have them scrutinize everything we're doing." By the time the upgrade is finished, they will have installed sprinkler systems and fire-rated walls and doors. As part of the upgrade, they are building new retention areas. The company also is writing a grant that will provide for an OSHA-certified instructor to train the entire staff on safety issues. "If anyone sees anything wrong," he says, "we'll know how to react to it." Three of the staff also are receiving training to become Red Cross first responders.

Chapter 4. Biodiesel Quality Standards

This chapter discusses the best practices used to ensure that consumers are purchasing quality biodiesel at the pump. Then, it explains why some biodiesel is of poor quality. In addition, it details how the use of biodiesel can affect car warranties. Finally, it offers a critique of off-the-shelf reactor designs.

Standards Used to Ensure Quality Biodiesel

ASTM D 6751 [Please refer to ASTM for updates to the biodiesel standard]

All commercially produced biodiesel that is intended to be blended at B20 or lower must meet a quality standard known as ASTM D 6751. (Petroleum diesel must meet its own quality standard.) This standard was developed by the American Society of Testing and Materials (ASTM), which is an international standards development organization. To determine if their biodiesel meets the ASTM D 6751 quality standard, the producer must submit a fuel sample for testing – this process costs about $1,000 and up and can be prohibitive for small producers. However, it is important because we want to ensure that expensive engines and fuel systems are not being subjected to substandard fuel.

Several government bodies are responsible for ensuring that biodiesel produced commercially meets the ASTM D 6751 quality standard. Some of these are discussed below.

Internal Revenue Service (federal) – Employs approximately 130 fuel compliance officers that inspect fuel quality nationwide. They monitor blenders who claim a tax credit on the biodiesel. Blenders are individuals or businesses that mix biodiesel with traditional petroleum diesel at any percentage. The tax credit (established as part of the American Jobs Creation Act of 2004) applies to both those who wish to resell their fuel and those who use it themselves. Eligible parties may receive one cent per percentage of biodiesel blend per gallon (thus, a blend of 20% would receive a 20 cents per gallon tax credit). The blender, however, must receive written certification that the biodiesel meets ASTM D 6751 standards.

Penalty: Up to $100,000 for an individual, $500,000 for a corporation, plus possible jail time if the producer or the blender knowingly signs any document that contains false information. For instance, if the producer falsely claims that the biodiesel meets certifications, then the producer may be penalized. A blender also can be penalized by, for example, falsifying documents stating that the biodiesel meets ASTM certifications.

Environmental Protection Agency (federal) – Biodiesel must be registered with the EPA and meet the standards set forth by the Clean Air Act. This only applies to those selling biodiesel, not biodiesel made for personal use. However, private use producers may need to obtain air quality permits from their states' environmental agency depending on the volume made. Chapter 4 discusses this further, but you can contact your state agency for additional, up-to-date guidelines. ASTM D 6751 is used as the standard to meet the provisions of the Clean Air Act. EPA officials have the right to inspect biodiesel plants, refineries, importers, distributors, and others and to conduct tests on the fuel. Motor fuel regulation falls under the EPA's Office of Enforcement and Compliance Assurance and by the Certification and Compliance Division of the agency's Office of Transportation and Air Quality.

State Agencies – Much of the enforcement responsibility lies with the states, although state enforcement levels vary greatly. Some states regulate all fuels, including biodiesel, while some do not. The agencies that regulate fuels are usually the state's Division of Weights of Measures, Department of Agriculture, or a similar agency. Biodiesel, being a relatively new fuel, has yet to be recognized by many states as needing regulation; however, a growing number of states are catching on. The National Biodiesel Board has compiled a list of regulatory information for each state, and this information is updated regularly. To find information on your particular state, visit the National Biodiesel Board's web site at http://biodiesel.org/resources/fuelqualityguide/states.aspx.

Tests included in ASTM D 6751

Next, we will take a look at the standards that producers have to meet.

The following tests are included in the ASTM D 6751 standard:

- Flash Point (D93)
- Water and Sediment (2709)
- Viscosity (D445)
- Sulfated Ash (D874)
- Sulfur (D5453)
- Copper Strip Corrosion (D130)
- Cetane (D613)
- Cloud Point (D2500)
- Carbon Residue (D4530)
- Acid Number (D664)
- Free Glycerin (D6584)
- Total Glycerin (D6584)
- Phosphorus Content (D4951)
- Reduced Pressure Distillation (D1160)

Each of these tests is discussed briefly below (Bently Tribology Services 2007).

1. Flash Point – Specification: 130°C minimum: The flash point is the temperature at which the biodiesel burns. A flash point below 130°C indicates excess methanol, often resulting from inadequate methanol recovery. Inadequate washing also will lead to excess methanol. A low flash point makes the biodiesel more like gasoline than diesel fuel.

2. Water and Sediment – Specification: 0.05 % volume max: The presence of water can promote microbial growth and corrode fuel system components. Microbial growth eventually can lead to filter plugging and engine failure. Unlike petroleum diesel, biodiesel is much more likely to absorb water. Care must be taken to keep it water-free at every stage from production to the fuel tank. This test allows for the presence of a maximum amount of 0.05 % water by volume. Biodiesel will retain up to 1500 ppm of water.

3. Viscosity – Specification: 1.9-6.0 mm2/s: Viscosity is a function of saturated fat content. Although the viscosity must be adequate for proper lubrication and to prevent leakage in seals, very high viscosity also is undesirable because it raises injection pressures, thereby reducing equipment life. As biodiesel ages, its viscosity increases. Thus, measuring viscosity is useful in determining the oxidation of the fuel.

4. Sulfated Ash – 0.02 %/wt max: When testing the fuel, a sample is completely burned to remove all organic compounds. An ash value >0.02%/wt indicates residual catalyst in the fuel.

5. Sulfur – Specification: <15 ppm: Recycled feedstocks can have excessive levels of sulfur and may not meet requirements. Low sulfur content is necessary to reduce harmful sulfur dioxide emissions. B100 is virtually sulfur free.

6. Copper strip corrosion – Specification: No. 3 max: This test addresses a fuel's potential to damage the fuel system's copper

components. Fuel made from an oil feedstock with a high free fatty acid content, such as waste cooking oil, is more likely to corrode copper parts.

7. Cetane – Specification min 47: This test relates to the combustion properties of the fuel in the ignition process. Specifically, it measures the fuel's ability to ignite automatically under compression. Cetane is to diesel fuel what octane is to gasoline. A low cetane or octane rating results in the "knocking" sound that cars are known to make. Soy oil produces a cetane around 50, and the cetane that yellow grease produces can be as high as 65.

8. Cloud point – Specification: Must be reported: There is no ASTM requirement for this test, which is a function of saturated fat content; however, it must be low enough for customer requirements and climate. If animal fats are used as the feedstock for biodiesel, they are likely to cloud at room temperature because they contain more saturated fats. Vegetable oils (e.g., soybeans) have a lower saturated fat content and do not cloud as easily. Cloud point is important because, when it is cold enough for a fuel to cloud, it might not flow well through the vehicle's fuel lines. Cloud point of vegetable oil-based biodiesel will vary depending on the feedstock oil, with rapeseed/canola oil displaying superior cold weather characteristics among commonly available feedstocks.

9. Carbon residue – Specification: 0.05 %/wt max: This test measures the carbon deposit-forming tendencies of the fuel. High levels of glycerin could cause this test to fail.

10. Acid number – Specification: 0.80mg KOH/g: This test measures the presence of free fatty acids (FFAs). FFAs require a high concentration of potassium hydroxide (KOH) or sodium hydroxide (NaOH), which are the catalysts used for making biodiesel. Typically, waste cooking oil has a higher FFA content than do virgin seed oils. As biodiesel ages, the double bonds in the unsaturated esters oxidize, causing a rancid smell and forming aldehydes and keotones, both

of which are pollutants. This oxidation also can corrode auto parts. The acid value is about twice the value (%) of FFA present.

11. Free Glycerin – Must be <0.02%/wt.: Glycerin can cause fuel separation and engine carbon deposits. During biodiesel production, free glycerin is removed through settling and proper water washing. Free glycerin that is left in the biodiesel will settle to the bottom of storage tanks/fuel tanks, attract water, and promote algal growth.

12. Total Glycerin – Must be <0.24%/wt: This test requires a gas chromatograph. This requires a conversion rate of triglycerides to esters >98%. Often, a higher level of total glycerin results from an incomplete reaction in making the fuel. Consistently using biodiesel with a high level of total glycerin will damage fuel injector parts and lead to engine failure.

13. Phosphorus content – Specification: 0.001%/wt: Phosphorus is measured because it will destroy catalytic converters. When using virgin oil, the phosphorus should be removed in the processing of the beans.

14. Reduced pressure distillation – 360°C max: This test ensures that components with high boiling points (e.g., used motor oil) have not been added. The atmospheric boiling point of biodiesel generally falls between 330° and 357° Celsius.

BQ-9000

Although ASTM D 6751 is a mandatory standard for commercial biodiesel, the National Biodiesel Board has created an additional, voluntary quality standard known as BQ-9000. The National Biodiesel Board established the National Biodiesel Accreditation Program to provide accreditation for this quality standard. This new standard is being embraced by some US and Canadian producers, marketers, and distributors as a means to address quality issues beyond simply the finished product.

BQ-9000 combines the biodiesel quality criteria of ASTM D 6751 with a more holistic quality program that addresses more of the biodiesel life cycle – not just the finished product. As such, BQ-9000 addresses sampling, testing, storage, blending, shipping, distribution, and fuel management.

There are two types of accreditation – accredited producer and certified marketer. Accredited producers are judged based on testing, sampling, storage, retaining samples, and shipping. Certified marketers are those companies that sell B100 or blends. They must handle biodiesel properly.

Accreditation requires companies to undergo an independent audit of their quality control processes. Companies must fill out an application and pay $1,000 for the initial application fee and $2,000 for the initial audit fee. The application materials are reviewed, and an auditor is assigned and scheduled. Companies that pass the audit are granted accreditation for three years. Surveillance audits are conducted annually.

Problems with Biodiesel Purchased at the Pump

Optimally, fuel producers would pass the quality standards every time. It is unlawful to distribute biodiesel that does not meet the ASTM specifications. Most consumers trust government agencies to enforce the standards strictly to ensure that the product that they put into their fuel tanks will not damage their vehicles. Engine manufacturers also have an interest in fuel quality because they do not want off-spec fuel causing problems that would be covered under warranty. All of this finally brings us to the question: does the fuel purchased at the pump actually meet the ASTM standards?

A study performed in 2004 by the National Renewable Energy Laboratory and Magellan Midstream Partners LLC tried to shed some light on the quality of biodiesel fuel being sold at the pumps (McCormick et al. 2005). The survey consisted of fifty fleets nationwide that represented all major US biodiesel blenders and producers (>1 million gallons per year). The study took samples from biodiesel blenders and distributors nationwide that were obtained from a list on the National Biodiesel Board's web site. Overall, the researchers collected twenty-seven samples of B100 and fifty samples of a B20 blend. Samples were collected during the 2004 calendar year and analyzed at various laboratories.

An initial survey was performed by phone by which the interviewer asked a number of sellers a series of questions about quality issues pertaining to biodiesel. It was discovered that the biodiesel sellers believe that the responsibility for ensuring fuel quality lies with the producer from whom the seller purchases the fuel. Little inspection is performed on the fuel after it leaves the producer, and any inspection that is performed is largely visual. In general, those surveyed believed that the product met quality requirements even though most did not receive analytical reports with each shipment of biodiesel. After the sampling and testing portion of the survey was performed, a follow-up phone survey was conducted about possible problems that the distributors have encountered (McCormick et al. 2005).

The samples of B100 were compared to the ASTM D 6751 specifications. Out of the twenty-seven samples, four failed at least one of the standards, which represented about 15% of the total. Next, the B20 blends were investigated. The studies were performed in late summer and, therefore, do not include winter operation. Because there is not, yet, a standard for a B20 blend (currently, there is a standard for B100 and a standard for petroleum diesel, which is used in blends to make B20), a list of tests was proposed by the National Biodiesel Board Fleet Evaluation Team. One recommendation was that a test be developed to include oxidation stability, which is an important factor in the growth of the biodiesel market.

One cause of concern is more related to blending processes

than production quality. A test was performed to determine if the entire B20 sample did, in fact, contain 20% biodiesel. Out of fifty samples, only thirty-two were close enough to 20% biodiesel to be considered B20 (between 18% and 22% biodiesel). The other eighteen samples ranged between 7% and 98%.

These results show a cause for serious concern in blending practices. Most blenders employ splash blending in which they splash the biodiesel on top of the diesel fuel in a truck or tank. Biodiesel is denser than diesel, so, if it is not mixed properly, it will settle to the bottom of the tank. Pumping from the bottom of the tank will yield a high concentration of biodiesel in this case. This is a problem because high concentrations of biodiesel can be corrosive to rubber engine parts. A lower concentration of biodiesel will not necessarily be bad for the car, but it misleads consumers, who believe that they are getting a different blend. The industry has recognized this problem and is working to improve its blending techniques.

How Biodiesel Affects Car Warranties

Every automotive and engine manufacturer requires testing and certification of transportation fuels to minimize the risk of damage and related warranty costs. Car and engine manufacturers place warranties on their products for "materials and workmanship." If something goes wrong with the engine within the stated warranty period and it is the result of a problem with how the engine was manufactured or the materials were used, the company will accept responsibility for the error and fix it free of charge. Car and engine manufacturers do not place warranties on fuel, but they do specify what fuels should be used in their engines. According to the National Biodiesel Board, if engine problems are caused by the fuel used, this is not covered by the engine warranty. Instead, such problems would be the responsibility of the fuel producer. Nonetheless, some engine warranties will be voided if a different fuel is used than the one recommended.

The diesel fuel sold at typical gas station pumps is suitable for most diesel vehicles. Using this fuel will not void a warranty. Biodiesel, however, has certain limitations on usage that vary based on the car manufacturer. Most manufacturers allow biodiesel blends to be used in the engines without voiding the warranty, but they vary on the level of blending. Below is a list of some major car and diesel engine manufacturers and the maximum level of biodiesel blends that they allow in their vehicles without voiding the warranty. If you are in the market for a diesel vehicle, be sure to check with the manufacturers frequently for possible updates to these numbers.

Ford Motor Company – B5
General Motors – B5
Toyota – B5
Volvo – B5
Volkswagen – B5
Mercedes-Benz – B5
Caterpillar – B5 and B30 in some engines
John Deere – B5
Kubota Tractor – B5
Ford-New Holland Tractor – B20
Daimler-Chrysler – B5, B20 in 2007 Dodge Ram (National Biodiesel Board 2007, Standards and Warranties)

Although these are relatively low blends, as engine manufacturers gain more confidence in biodiesel, they are likely to increase the allowances to B20 in the near future. Many are waiting on a much-anticipated nationwide ASTM standard for B20. To raise support for B20 levels, the National Biodiesel Board has teamed up with diesel engine, fuel injection, and vehicle companies to form the B20 Fleet Evaluation Team. The results have been very positive. Beginning in 2003, the National Biodiesel Board (NBB) commissioned a survey of over 50,000 diesel vehicles in fifty-three different fleets. Ninety-six percent of the biodiesel users stated that they would recommend using B20 in other fleets. Over the past few years, there have been over fifty million problem-free miles driven on blends of B20 (NBB 2005), and the major automotive companies finally are gaining assurance in higher biodiesel blends.

Consider the following statement from the GM fleet account executive Brad Beauchamp: "While we have seen no trouble in using B20, we are waiting to change our warranties until the fuel quality is consistent enough... so that we feel comfortable that it won't damage the engine" (Anonymous 2004). Biodiesel offers additional power and lower emissions than standard diesel according to Beauchamp. He said that the warranties for all GM vehicles with diesel engines would be updated to allow for the use of B20 as soon as an official ASTM B20 standard is passed. For now, however, manufacturers are hesitant to increase the allowable biodiesel blend for fear of adverse effects to their engines and engine parts, which has been the catalyst for the development of the B20 ASTM specification.

There is much overlap in vehicle manufacturers' lists of potential problems that could result from high biodiesel blends. Typically, manufacturers might state that blending biodiesel fuel above a 5% concentration could have some adverse affects on the engine, such as those identified below.

- Storage is a problem because of a higher-than normal risk of microbial contamination due to water absorption and rapidly decreasing oxidation stability (sunlight affects oxidation stability), which creates insoluble gums and sediment deposits.

• Being hydroscopic, the fuel tends to have increased water content, which increases the risk of corrosion. A hygroscopic substance is one that rapidly attracts water from the atmosphere.

• Biodiesel is an effective solvent (e.g., it can strip paint) and will tend to loosen deposits in the bottom of fuel tanks of vehicles previously run on petroleum diesel.

• The methyl esters in biodiesel fuel might break down the seals and composite materials used in vehicle fuel systems.

• Biodiesel leakage through seals and hoses can induce corrosion on rubber parts.

• Biodiesel can corrode fuel injection equipment.

• Injector nozzles can become choked/blocked, leading to poor atomization of fuel.

• Internal injection system components can become lacquered/seized.

• Sludge and sediments accumulate. (John Deere 2005; Ford 2005)

FIE Manufacturers is a consortium of fuel injection equipment manufacturers, including Delphi, Bosch, Siemens VDO, Denso, and Stanadyne. While they support biodiesel development, they are concerned about fuel quality. The following have been listed as contributing to substandard fuel: free methanol; water; free glycerine; mono, di-, and triglycerides; free fatty acids; total solid impurity level; alkali/alkaline earth metals; and oxidation stability. Although they do support biodiesel blends up to five percent, the FIE Manufacturers have stated publicly that ASTM D 6751 addresses neither oxidation stability nor safeguards to protect the quality of biodiesel blends.

Factors Affecting
Biodiesel Quality

There are many independent variables that affect the quality of biodiesel before, during, and after the production process. Just as important as ensuring quality control in the production process is quality control after the final product is produced. Each time biodiesel is transferred, pumped, stored, and blended, the possibility of contamination presents itself. Unfortunately, the only tried and true method of ensuring that biodiesel remains free of these pests is to develop proper handling, storage, monitoring, and remedial measures. The NREL Biodiesel Handling and Use Guidelines are an excellent starting point for someone hoping to understand proper biodiesel handling, storage, and distribution methods. It is available online at http://www.nrel.gov/vehiclesandfuels/npbf/feature_guidelines.html.

Feedstocks

Rendered animal fats, vegetable oils, and plant oils differ from petroleum diesel fuel mainly in viscosity (Kemp 2006). These materials are of a chemical structure consisting of triglycerides, which are chemical compounds formed from one molecule of glycerol and three fatty acids. Vegetable oils contain varying amounts of fatty acids, which are chains of hydrocarbons differing in carbon length. Characteristics, such as oxidation stability and cloud point, are determined from fatty acid composition.

Free fatty acids differ from the fatty acid chains. Free fatty acids, such as triglycerides, are not bonded to the parent oil molecules. This presents problems in producing quality biodiesel. Generally, feedstocks with higher free fatty acids (FFAs) present the most challenges in biodiesel production. Because of cost, higher FFA feedstock, such as fryer or waste oil, is typically used by home brewers.

The general category of fryer oils and fats (or waste oil) comprises vegetable oils, semi- or fully hydrogenated vegetable oil, and animal fats, such as lard and tallow (Kemp 2006). After repeated use, fryer oils and fats break down and, because of increasing FFA content, eventually will become unsuitable for cooking.

Typically, fryer oil unsuitable for cooking is recycled by rendering companies. The rendering company provides the restaurant with a storage tank and periodically empties the tank after which the oil is processed into yellow grease. This is a big industry, and it has been estimated that the recycling and reprocessing of waste oils and fats generates 2.75 billion pounds of yellow grease annually (Kemp 2006, 130). To say that used oils and fats from restaurants are a "waste product" is an exaggeration since an entire industry has been built around the recycling and reprocessing of it into yellow grease.

Waste oils and fats also are deposited into sewer systems as a result of washing after cooking. Small amounts of oils and fats are released and can damage municipal waste water systems and be very expensive to repair. Therefore, grease traps are required to catch these deposits, which commonly are referred to collectively as "trap grease." Trap grease can be broken down through the use of enzymes or chemicals but often are removed by a vacuum service. The resulting contaminated water often is landfilled and is called "brown grease" (Kemp 2006).

The lowest free fatty acid levels are found in refined vegetable oils, which generally contain less than one percent FFAs. Waste oil and fats from restaurants contain between two and seven percent FFAs while yellow grease and animal fats, such as beef tallow and lard, can contain as much as thirty percent FFAs. The highest levels of FFAs are seen in brown grease (typically above thirty percent), which explains why few biodiesel producers are utilizing this feedstock despite its very low cost (typically free) (Kemp 2006, 108). Even at a relatively low level of two percent FFAs, the chance of forming soaps and causing emulsion formation during washing is increased greatly (Kemp 2006, 397).

Free fatty acid levels are determined using a process called titration. Based on this test, additional sodium or potassium hydroxide is used to neutralize FFAs.

Waste oil quality varies widely – oftentimes even at the same restaurant. Some best practices are listed below that might help to ensure better quality waste oil, in turn enhancing the prospect of producing quality biodiesel:

• Ask if you can obtain waste product directly from the fryer (as opposed to a recycling storage unit or rendering container).

• Ask if the restaurant can rinse the fryer after they dump oil into the containers that you provide.

• Supply the restaurant owner with a box of 3M™ Shortening Monitor test strips, which are used to determine the quality and life of fryer oil.

• Test (and document) each container of used fryer oil to determine the FFA level. Mix between the selections to obtain an average FFA level if necessary.

Dealing with high FFA feedstock, such as brown grease, is an art that Philadelphia Fry-O-Diesel seems to have perfected. Pretreatment, known as direct acid esterification, is used with high FFA feedstock prior to transesterification.

Water, Microbial Growth, and Oxidative Stability

Because water, microbial growth, and contaminants can enter the supply chain at any time, quality testing is important at every step in the fuel distribution chain, including at storage and retail sales facilities. Again, possibilities for contamination present themselves many times throughout the supply chain; the only way to avoid a degradation of quality is to develop proper handling, storage, monitoring, and remedial methods.

Biodiesel's biodegradability properties can create long-term storage and fuel stability problems. Biodiesel will deteriorate at a more rapid pace when exposed to sunlight, elevated temperatures, or oxygen or when it is in contact with non-ferrous metals, such as aluminum, tin, copper, zinc, and brass (Kemp 2006, 91). The result is fuel thickening.

Water in biodiesel tends to create an ideal place for microorganisms to grow. Since biodiesel is hydroscopic, it attracts water more readily, creating a more difficult storage environment than its petroleum diesel counterpart. In fact, a study done by the University of Idaho found that biodiesel absorbed fifteen to twenty-five times more moisture than petroleum diesel did at the same temperatures (University of Idaho 2006).

To remove excess water – thereby improving chances of meeting the ASTM specification for water and sediment as well as your biodiesel's oxidation stability and decreasing the change for microbial growth – incorporate a drying step into your biodiesel production. William Kemp's drying tank, featured in his book *Biodiesel Basics and Beyond*, incorporates an electric water heater, circulating pump, heated air blower, and filter arrangement.

Biodiesel Handling and Storage
Biodiesel Blending

If you intend to blend biodiesel with petroleum diesel, it is important to understand the proper procedures so that the desired blend of biodiesel is achieved. One of this book's co-authors, Chelsea Jenkins, works for the Hampton Roads Clean Cities Coalition and has written this about biodiesel blending: Biodiesel is blended into diesel fuel via three primary means:

1. B100 is splash blended with diesel fuel by the distributor at the time that the delivery truck is loaded. Blending only at the end-user's tank is not recommended unless thorough precautions are taken to ensure adequate mixing.

2. Pre-blended (via a variety of means) by a jobber or distribution company in bulk storage tanks and offered for sale as a finished blend, often B20 or B2.

3. Blended at a petroleum terminal with automated equipment. This method, though not offered yet in many locations, ensures complete blending and reduces handling costs for distributors.

Splash Blending

Splash blending refers to an operation in which the biodiesel and diesel fuel are loaded into a common vessel from the same or separate sources with some mixing occurring as the fuels are pumped into a common tank. The vessel is usually an individual fuel tank on a fuel delivery truck (or a drum or tote). Once the fuels are splash-blended onto a delivery truck, additional mixing occurs as the truck travels to customers' locations. This approach can be successful if agitation is adequate, but is not a recommended practice, especially in cold weather. *Little or no mixing can occur if biodiesel is loaded first into an empty delivery truck tank on a very cold day.*

Pre Blending

The biodiesel and diesel fuel are blended by the distributor and bulk-stored in that blended form before being loaded onto a delivery truck.

In-Line (Injection) Blending

The biodiesel is added to a stream of diesel fuel as it travels through a pipe or hose in such a way that the biodiesel and diesel fuel become thoroughly mixed by the turbulent movement through the pipe – and by the additional mixing that occurs as the fuels enter the receiving vessel. The biodiesel is added slowly and continuously into the moving stream of diesel fuel via a smaller line inserted in a "Y" in a larger pipe; otherwise, the biodiesel can be added in a small slug or in pulsed quantities spread evenly throughout the time that the petroleum diesel is being loaded. This is similar to the way in which most additives are blended into diesel fuel today and is most commonly used at pipeline terminals and racks. Some distributors carry B100 and petroleum diesel in separate truck compartments and blend the two fuels with separately metered pumps operating simultaneously at high speeds. These methods offer superior consistency and lower operational costs.

Tests to determine if fuel is thoroughly mixed:

1. A top, middle, and bottom sample of the tank contents (see ASTM D405712 for the proper way to take a representative sample of a tank's contents) can be taken and analyzed for the percent of biodiesel using infra-red spectroscopy or by measuring the specific gravity or density. See www.biodiesel.org for more details.

2. Put the samples from the three layers in a freezer with a thermometer and check every five minutes until the fuel in one of the samples begins to crystallize. Record that temperature. Then, check every couple of minutes or so until all three samples show crystallization. Compare the crystallization temperatures on all three samples; they should be within 5-6°F (3°C). If not, the fuel will require agitation to mix thoroughly.

Cold Weather Blending Considerations

Blending and storing biodiesel in cold weather can be difficult and requires extra precaution and attention to detail. An important step is to identify cold flow properties of the fuel that you have made or are purchasing.

To successfully blend biodiesel in cold weather, follow the tips below:

• Identify the cold flow protection desired (this will depend on the area in which you are located).

• Obtain a generic fuel with the lowest "temperature operability value" possible (cloud, cold filter plugging point, and pour point).

• Use a proven additive or kerosene to reduce the generic fuels operability value to levels low enough to accommodate the percentage of biodiesel blend desired. A twenty percent biodiesel blend produced from soy methyl ester will reduce your cold flow values by 10°F in some cases, and waste oil used as a feedstock will reduce cold flow values to a higher degree.

• Blend biodiesel into diesel fuel and test the product for operability values. When splash blending is your only option, the hotter the biodiesel, the more likely you will have a successful blending experience. To eliminate the possibility of the biodiesel flash freezing when introduced into a cold aluminum tank truck, successful blenders have heated the fuel to temperatures in excess of 100°F. (An NBB study concluded that successful B2 blends were made when the biodiesel was 10°F above its cloud point (NBB 2005b)).

The fuels will remain blended once you have blended biodiesel into the diesel fuel or heating oil following the general blending principals discussed above.

Critique of Off-the-shelf Reactor Designs

"Home brewers" are very resourceful when it comes to constructing their own biodiesel production facility, often using surplus tanks, pumps, valves, and buckets to embark cheaply on their endeavor.

For those who want some guidance, there is no shortage of literature on the web about constructing your own processor. There also are commercial-off-the-shelf processors available from companies for small-scale production. These are almost exclusively batch processors, which are significantly less expensive and complicated but not as automated as their continuous flow counterparts. The batch processors, as their name implies, work on a batch-by-batch system, which is limited by the size of the processor. These are more labor-intensive and require a standard, precise preparation of ingredients prior to each batch being made. Continuous flow processors, on the other hand, produce biodiesel constantly with very little downtime. They only need a constant supply of oil, alcohol, and catalyst to be fed into the system and will produce biodiesel with high automation; however, they are more costly than are the batch processors.

Plenty of forums exist online (e.g., biodieselnow.com, biodiesel-community.org, biodiesel.infopop.cc) for the ever-increasing number of home brewers who want to make their own biodiesel but need direction on getting started with a processor. There are many designs and products out there, but there has been little side-by-side comparison of small-scale processors. Would-be biodiesel producers are advised to sift carefully through the wealth of information on the Internet, always keeping in mind that safety and fuel quality are crucial responsibilities. Look for information backed up by research and popular support of the biodiesel community.

Researchers at James Madison University in Harrisonburg, Virginia, and Virginia Polytechnic in Blacksburg, Virginia, joined efforts with the Hampton Roads Clean Cities Coalition to conduct tests on four commercially available small-scale processors. Two of the processors studied were commercial-off-the-shelf processors (COTSP): the FuelMeister and the Biodieselgear 60 (each costs between $2,000 and $3,000). A third was a very popular "open-source" processor known as the Appleseed ($800-$1,500), and plans for that processor can be found online. The final one was a James Madison University (JMU) processor, known as JMU I, that was built by university students (about $2,400). A detailed discussion of this study can be found at Brodrick et al. (2005).

Some characteristics of the processors are shown below with relevant comments and recommendations.

Heating

FuelMeister: Electric barrel wrap on supply barrel
Biodieselgear 60: Hyrdronic piping loop in main tank
Appleseed: Electric insertion heater in main tank
JMU I: Electric insertion heater in recirculation loop

• The band heater on the FuelMeister can be used only in a 55-gallon metal drum of waste oil, which was an inconvenience because it was difficult to control heat dispersion in the drum, and the hot band was difficult to adjust.

• The heater on the Biodieselgear 60 heated more evenly, but the thermostat gives the temperature of the recalculating water rather than the oil.

• The Appleseed and JMU processors heated evenly with insertion heaters and allowed direct temperature monitoring with thermostats.

Method of Mixing Methoxide

When the methanol (alcohol) and potassium hydroxide (catalyst) are mixed, they form methoxide.

FuelMeister: Performed in 16-gallon mixing tank using a manual pump handle; draws methanol from large drum
Biodieselgear 60: Mixed in polyethylene tank by rocking the tank back and forth

Appleseed: Mixed in a sealed polyethelene carboy container
JMU I: Prepared in a container under a fume hood

• All of the systems except for the JMU I had a closed system for mixing the methoxide. The JMU I was able to be mixed in an open container because there was access to a fume hood. Otherwise, this process must be done in a closed container.
• None of the methods was easy to control. Constant monitoring was required, and there was no visual way to tell if the catalyst was mixed adequately.

Type of Processor Tank
FuelMeister: 55-gallon plastic cone bottom tank
Biodieselgear 60: 60-gallon sealed polyethylene cone bottom tank on steel frame
Appleseed: 50-gallon electric hot water heater
JMU I: Two 15-gallon plastic cone bottom tanks

• Traditionally, plastic tanks have been favored because they are inexpensive, are easy to drain and clean, and allow for visual inspection. However, high temperatures can damage these tanks, and many have been known to leak after long-term use. In some instances, fires have been started.
• The Appleseed tank is the most desirable because the tank is insulated and built to withstand high temperatures. It contains a pressure and temperature relief valve along with a reliable thermostat. A clear hose is included to allow for visual inspection. Used water heater tanks often are recovered free from plumbers' salvage yards. The Appleseed also has the advantage that it can be modified to accommodate methanol recovery equipment, which improves overall sustainability and reduces the complication of byproduct handling.

Other comments and recommendations are as follows:

- None of the reactors had a way to pre-filter the waste cooking oil before being transferred to the processor. Filtering is highly recommended and should be done prior to heating the oil to remove solid particles. Simple pre-filtration can be added to any reactor using readily available bucket filters or other off-the-shelf components.

- The method of mixing methoxide needs to be improved on all systems. An automated mixer would be much more convenient.

- A methoxide injector is recommended to add the methoxide to the main processor. If the catalyst is mixed in an open container, it must be done under a fume hood.

- Stainless steel or black malleable steel should be used for tubing, valves, and other parts. PVC and plastic tubing degrade over time. (Brass ball valves commonly are used without major problems. Aluminum must be avoided due to the tendency to react with NaOH or KOH catalysts.)

- Install a system to limit the amount of wash water added to prevent overfilling the processor tank.

In conclusion, what is momentously more important than the type of processor used are the techniques employed by the individual. After a few minutes of web surfing, one will be exposed to a plethora of information from "experts" who claim to know the key to making quality fuel. These forums, group pages, and even processor manufacturers tout how simple it is, that everyone can do it, and how much money it can save on fuel. As one experienced home brewer put it, "They make it sound as easy and cheap as those tabloid ads do for losing twenty pounds."

However, it is neither easy nor cheap. Plenty of people have spent thousands of dollars on commercial units only to make poor quality fuel that damages their vehicles. In addition, many people do not follow the proper disposal methods and end up using environmentally destructive practices. What starts out as being a seemingly environmentally friendly way to save money on fuel becomes a frustrating loss of time and money. Poor quality biodiesel gives it a bad reputation and hampers its growth as a viable alternative fuel.

This is not meant to discourage people from making their own fuel. We are strongly in favor of locally produced, clean-burning fuel. There are plenty of people who do this at home or in their communities and enjoy great success. But there also are many who are not doing it correctly. Take the time to do proper research and do not believe everything that you read on the Internet, especially if it claims that making biodiesel is easy and provides shortcuts in the process. Creating ASTM-quality biodiesel on a small-scale basis is difficult and takes a fair amount of technical know-how. There is no requirement for home brewers to make ASTM quality fuel for their own use; however, by not using quality fuel, be aware that you put your vehicle at risk. This might appear to be a fair tradeoff for some people who own older cars and would like to make cheaper fuel. Many people cannot or do not want to spend the $1,000 to test their fuel, but caution and care go a long way.

When doing research on making biodiesel, try to find credible sources. Look for advice from people who have been known to make ASTM quality fuel. Robert Miller, whose case study is featured below, is an expert who produces ASTM quality fuel. It is possible to produce quality biodiesel, but the necessary due diligence needs to be taken to do it correctly, safely, and without harming the environment that you intend to help.

Case Study: Robert Miller, Madison, Virginia

Robert Miller began producing biodiesel from waste vegetable oil in the fall of 2004. He has designed and built four versions of a

42-gallon (1 barrel) batch processor utilizing industrial techniques and materials gleaned from his experience as the designer, builder, and operator of a farm-scale ethanol plant in the 1980s and 1990s. After researching the various biodiesel production methods, he settled on a double reaction process, which he says can produce ASTM quality fuel consistently with little worry. He is very particular about the recovery of the excess methanol and the partial refinement of the glycerol byproduct.

Robert chose to produce biodiesel for his farm tractor and a recently purchased family car, a Jetta TDI wagon, as a response to the low fuel economy of gasoline cars on the market and a desire to break his family's dependence on non-renewable fuel.

The process begins with used vegetable oil picked up from the local Chinese restaurant, which consistently has an acid value below 3.5. He says, "Why struggle with oil from a local source with high acid values if you can get oil with lower values just as easily? I did that (struggled) for a time and ran an acid esterification stage until I realized I was mixing high FFA oil with low FFA from local restaurants." He adds, "Don't be in a hurry to make your fuel, take your time and do it right."

Since Robert recovers the excess methanol from his fuel using a vacuum pump before washing, he uses the vacuum pump in his reactor design to transfer all materials to the reactor. A reactor operating under vacuum is safer than one at atmospheric pressure or under any pressure; any potential vapor leaks enter into the reactor rather than leaving the reactor, which would occur with a reactor under pressure.

The process works like this:

1. Pull a vacuum on the reactor, and open the oil suction line.

2. Fill the reactor with oil to the proper level. Then, turn off the vacuum pump, and close the oil valve.

3. Turn on the heater and recirculating pump.

4. Heat the oil to 130 degrees Fahrenheit.

5. Turn on the vacuum pump to suck in sixty percent of the pre-mixed methoxide.

6. Maintain the temperature at 130 degrees for one hour.

7. Turn off the heater and pump, and let the mixture settle overnight.

8. Pump out the settled glycerol.

9. Repeat steps 1 through 8 using the remaining forty percent of the methoxide.

10. Heat the fuel to 150 degrees Fahrenheit, and pull a vacuum until no more methanol vaporizes.

11. Pump the mixture into the wash tank, add Magnesol, and agitate it all for thirty minutes.

12. Allow the Magnesol to settle overnight or longer if desired.

13. Pump the mixture into the dispensing tank through a 5 micron filter.

14. Pump the mixture from the dispensing tank though a 1 micron filter into your vehicle.

Robert suggests a few simple tests to ensure that biodiesel producers have made biodiesel properly.

1. Take two half-quart samples of finished fuel. Re-react one, and use the other as a control; look for glycerol settling out of the re-reacted fuel after twenty-four hours.

2. Take a half-liter sample of finished fuel and add water; then, shake vigorously. Look for the speed of separation and cloudy water or the formation of an emulsion.

3. Test the tap water and the water fraction from test #2 above with litmus paper. Look for neutrality.

4. Place a sample in the refrigerator undisturbed for twenty-four hours. Look for haze or the settling of a white film on the bottom of the jar, which is likely to indicate unreacted mono and di-glycerides.

5. On an electric stove, gently heat a pan of finished fuel to see if any water or methanol boils out. Take care not to be burned by fuel erupting from vaporizing water or methanol.

Robert has submitted a sample of his fuel to ASTM for testing and found it in compliance. He says that when people learn the he makes his own biodiesel, they ask how much it costs to make a gallon of biodiesel. He tells them "I have it down to about $800 a gallon…"

His entire process is controlled by a computer except for the glycerol separation. Dealing with the glycerol is important because glycerol and water interfere with and impede the transesterification reaction. Robert says that he has a solution to this but has not applied it because he, along with his colleague, Mark Allen, have recently invented a device and a new process that has a patent pending.

The new process uses a device they have invented and is called a RASP. The acronym RASP stands for **R**eaction and **S**eparation **P**rocessor. This patent-pending device performs both the biodiesel reaction and the separation of the glycerol in a continuous process. In addition to being highly efficient, the RASP process also is very compact – a 225,000-gallon-per-year plant occupies a footprint of approximately thirty square feet. RASP Technologies, LLC, will be marketing these plants beginning in early 2008 to communities, universities, and companies that have access to a supply of virgin or used oil.

Robert says that his company, RASP Technologies, is very excited to be able to help the small community biodiesel producer to achieve its goals. "The RASP has the potential to make smaller scale community production of biodiesel fuel a reality without the high capitol or labor costs usually associated with sophisticated continuous production units." In addition to the current low FFA model RASP unit, RASP Technologies also is developing a RASP process that is capable of efficiently converting high FFA oil into ASTM quality fuel.

He says, "Having operated a small ethanol plant through the '90s, I see great potential in community based, farm co-op owned, biodiesel fuel production. The farmer can, and I believe, will, become an important producer of fuel as well as food and fiber. It is imperative that he does, as it will sustain our local economies and open space we all cherish." However, Robert suggests, no one should consider making biodiesel if they do not possess the capacity to safely recover and recycle the excess process methanol.

Chapter 5. Biodiesel Production Regulations & Permitting

This chapter lays out some of the regulatory and permitting challenges facing homebrew and small, non-industrial biodiesel producers. Though it does not discuss every regulation and permit required to produce biodiesel in various states, it does give the reader an overview of some of the regulatory and permitting issues that small-scale biodiesel producers might face.

It often is said that the difficult part of biodiesel production is not so much the actual production of biodiesel but, rather, the regulatory spider web of permitting, complying with codes and standards, and registering with various organizations and agencies covering a wide variety of areas. Homebrew operators sometimes find that their biggest challenge is in determining which agency they need to contact. Once they determine who to talk to and which agencies have jurisdiction over their operations, then begins the process of regulatory education since biodiesel is relatively new to many government agencies.

Homebrew operations tend to function in a regulatory grey area, which can be advantageous in some cases and prohibitive in others. Worst-case scenario: every agency believes that homebrew operations are under its jurisdiction. Some agencies that may need to be consulted include building code and zoning officials, planning commissions, federal and local environmental protection agencies, hazardous materials and waste management divisions, federal and state tax agencies (departments of revenue), departments of motor vehicles, landlords, insurance providers, federal and state departments of agriculture, water and sewage agencies, state and local fire marshals, and so on.

Each state and locality takes a different regulatory approach to homebrew and small-scale biodiesel production. Furthermore, commercial-scale production is regulated differently from homebrew or small-scale operations – although it sometimes does not feel that way to small biodiesel producers who are trying to get up and running on a limited budget.

Biodiesel operations also vary widely in both structure and operation. This is a blessing and a curse. In one regard, it would help if there was a precedent to follow with the steps laid out for each type of operation. Then, a future biodiesel organization would have a less frustrating, more predictable journey. On the other hand, the beauty of biodiesel is that this homegrown fuel can be produced at a local level using local resources and in very different areas of the country. Thus, an operation in urban California has very different community implications than an operation in rural Georgia and will involve many different variables.

Biodiesel production is affected by two forms of government policy – regulatory policy and tax and subsidy programs (Schumacher 2007). Often, the regulatory burden is closely correlated with the amount of process chemicals, waste products, and biodiesel stored on site as well as the method of production and storage. Although state and federal agencies tax biodiesel, incentives also are present to encourage development of the market.

The realm of biodiesel production is unusual in that it is not limited to those with chemical processing and manufacturing experience but also includes the homegrown fuel enthusiast. As Lyle Estill writes in his book *Biodiesel Power* (2005), "biodiesel has buzz," and buzz tends to attract the mass public. However, anyone serious about biodiesel must become familiar with newer concepts, such as "class IIIB combustible liquid," "40 C.F.R. Part 79," and "RCRA."

There are several areas related to the construction, design, and operation of biodiesel manufacturing that could be regulated, which means that some states might regulate a specific activity currently, some might not, and some might in the future. Many rules, regulations, and taxes at the federal, state, and local levels apply. For instance, there are biodiesel health, safety, and zoning ordinances that are similar to those that apply to fossil fuel production, distribution, sales, or other oleochemical operations.

The early stages of planning a biodiesel initiative may leave one confused about which regulatory issues affect homebrew, on-farm, and commercial small-scale production. The sections below address many of the major areas of interest for biodiesel operations. Note that operations on zoned agricultural land or farms tend to get a regulatory pass on everything from building permits to fire inspections. Therefore, small-scale producers are advised to stay on farm whenever they can because obtaining permits on farm is far easier than it is otherwise (Estill 2007b). Biodiesel groups also have tried teaming up with a university (see James Madison University's biodiesel program http://www.cisat.jmu.edu/biodiesel/ or the Biodiesel Co-op of Kalamazoo at http://kzoobiofuels.org/ Home.html) or municipality (US Department of Energy 2004). These tend to have compliance, health, and safety officers already on staff to help investigate individual needs, such as liability, the disposal of sidestreams, and complying with state safety regulations for biodiesel manufacturing. Furthermore, necessary permits might already exist and be on file.

Environmental Regulations and Biodiesel Production

State environmental agencies might have jurisdiction over several areas related to the construction, design, and operation of a biodiesel production process. For example, Ohio's Environmental

Protection Agency (2007) states that environmental compliance in biodiesel production includes:

1. "Air emissions from tanks, pumps and valves, fuel burning equipment and material handling;
2. Process wastewater discharges from boiler blow down, cleaning and other production activities;
3. Management and disposal of wastes;
4. Storm water contamination from material handling, on-site storage and facility construction activities; and
5. Spill prevention, planning and notification."

The subsections that follow discuss these areas in greater detail.

Air Pollution Requirements

By their very nature, air permits and regulations are extremely complex. The air pollution requirements that must be met for biodiesel production depend on how much air pollution will be emitted from the operation or facility. Most permits are required to be obtained prior to any construction or production activities, and often are accompanied by a permit to construct and operate, and include emissions limits, monitoring and operating conditions, and record keeping requirements. As a result, we strongly recommend acquainting yourself with relevant regulations and contacting the proper agencies with jurisdiction early in the planning process.

Methanol is a volatile organic compound and is what is known as a criteria air pollutant and a hazardous air pollutant. Under the Clean Air Act, the EPA established the National Ambient Air Quality Standards (NAAQS) for six common air pollutants. These pollutants include particle pollution, ground-level ozone, carbon monoxide, sulfur oxides, nitrogen oxides, and lead. These are called "criteria" air pollutants because the EPA regulates them by developing human health and environmentally based criteria. Hazardous or toxic air pollutants are believed to cause cancer, other serious health effects, or environmental damage (US EPA 2007a).

Methanol is the concern of most state environmental departments that regulate air pollution. Not only is methanol used in the process, but it also can remain in biodiesel byproducts, thereby creating further regulated substances. Methanol is not the only potential source of air pollution that might be regulated or need permits.

Fuel burning equipment that is used in biodiesel production can emit other harmful chemicals as well and might need to be regulated (Oregon Department of Environmental Quality 2006). Others can include boilers or process heaters, electrical generators, material storage tanks, reactors, separators, evaporator units, truck loading racks and/or drum filling operations, and so forth (State of Ohio Environmental Protection Agency 2007).

In many cases, equipment and process units that have low emissions will not require air permits. Exemptions for small-scale producers can often be obtained if the producer documents that equipment and process units do not emit higher levels than state regulations allow. Exemption for all emissions sources is achieved most easily with a closed-loop process. Despite the possibility of a full exemption, registration often is required to be accompanied by periodic inspection and annual emissions updates.

Although you might qualify for an exemption, you still might need to submit an application and obtain an exemption letter as well as documentation demonstrating your compliance.

For example, in New Hampshire, there is a concern over unintended, or "fugitive," methanol emissions. These are regulated as air pollutants under the NH Code of Administrative Rules, which requires biodiesel producers to demonstrate compliance with emission limitations for all of the toxic air pollutants that are regulated. Any producer that might exceed these emissions limits might need to obtain an air permit. The state's Department of Environmental Safety already has reviewed several plans pertaining to small-scale production (<1,000 gallons per day) and determined that they would be in compliance without restrictions and would not require an air permit. Nonetheless, such facilities still would need to obtain documentation that demonstrates their compliance with the state laws (New Hampshire Department of Environmental Services 2006).

The state of Oregon requires a potential biodiesel producer to submit a Notice of Intent to Construct and include an estimate of all air pollutant emissions. Emissions of a single criteria pollutant above one-half ton per year requires an Air Contaminant Discharge Permit. Emissions of a single Hazardous Air Pollutant over ten tons per annum requires a Title V Air Operating Permit (Oregon Department of Environmental Quality 2006).

Listed below are examples of state air pollution requirements for Virginia, Ohio, Oregon, and New Hampshire. This list is meant to serve as a starting point prior to contacting the proper authorities in your particular state/locality.

Virginia

Exemption levels are contained in 9 VAC 5-80-1320 of the Virginia Regulations for the Control and Abatement of Air Pollution. Below is a list of permits that the Virginia Department of Environmental Quality (VADEQ) has identified as potential required permits resulting from small-scale biodiesel production.

Air permits **are required** if:

Process equipment, such as reactors, separators, evaporators, etc., exceed emissions exemption levels.
   • Permit required if proposed facility produces more than 1,102 tons of glycerin per year, which is typically the three

million gallon or more per year producer. Federal EPA rules also apply.

Oil extraction equipment or oilseed handling equipment emissions exceed exemption levels
   • Chemical extraction is a special case. Contact VADEQ official.

Materials storage tanks and transfer operations exceed vapor pressure and storage volume requirements
   • "Permit required unless storage tank and transfer operation involve a petroleum liquid with a vapor pressure less than 1.5 pounds per square inch absolute under actual storage or transfer conditions and tank is 40,000 gallons or less" (VADEQ 2007, 5).

Boilers or process heaters exceed maximum heat input and use a regulated fuel
   • "Permit required unless fuel burning unit uses solid fuel with a maximum heat input of less than 1 million Btu per hour, uses liquid fuel (on spec distillate or residual fuel oil) or liquid and gaseous fuel with a maximum heat input of less than 10 million Btu per hour, or uses gaseous fuel with a maximum heat input of less than 30 million Btu per hour. Units larger than 10 million Btu per hour or larger are subject to the New Source Performance Standards for Small Industrial – Commercial – Institution Stem Generating Units (40 CFR 60, Subpart Dc)" (VADEQ 2007, 5).

Electrical generators are used for purposes other than emergency
   • Permit required if use exceeds 500 hours of operation per year and the unit has an engine other than a "gasoline engine

with aggregate rated brake (output) horsepower of less than 910 hp or a diesel engine with an aggregate rated brake (output) horsepower of less than 1,675 hp" (VADEQ 2007, 5).

Important Links—Virginia:

Air Regulations:
http://www.deq.virginia.gov/air/regulations/airregs.html

Exemption levels in 9 VAC 5-80-1320:
http://www.deq.virginia.gov/air/pdf/airregs/806.pdf

Form 7:
http://www.deq.virginia.gov/air/justforms.html

Regional Offices:
http://www.deq.virginia.gov/regions/homepage.html

Ohio

Typically, a *de minimus* source classification will be granted and permits will not be required if equipment and process units have low emissions and if it has been documented that air emission sources do not emit more than ten pounds of air pollutants per day. This exemption from air permitting requirements is most easily obtained by designing a closed-loop process.

Air permits **are required** if:

Process equipment, such as reactors, separators, evaporators, etc., exceed emissions exemption levels.
   • Permit required if proposed facility produces over 1,102 tons of glycerin per year, which typically comes from facilities

annually producing three million gallons or more per year. Federal EPA rules also apply.

Materials storage tanks and transfer operations exceed vapor pressure and storage volume requirements.
 • "Permit required unless tank has submerged fill and its capacity is less than 19,815 gallons; between 19,815 and 39,894 gallons and materials stored has a true vapor pressure less than 2.176 psia (Note: the vapor pressure of methanol exceeds 2.176 psia; or is greater than 39,894 gallons and material stored has a true vapor pressure less than .508 psia" (Ohio EPA 2007, 2).

Boilers or process heaters exceed maximum heat input and use a regulated fuel
 • Permit required if rating greater than 10 million BTU/hr and burns natural gas, distillate oil or liquid petroleum gas.

Electrical generators are used for purposes other than emergency
 • Permit required if use exceeds 500 hours of operation per year

"Truck loading racks and/or drum filling operations exceed de minimus levels" (Ohio EPA 2007, 2).
 • Permit required

Air permits for units usually are accompanied by a permit-to-install and a permit-to-operate and must be obtained prior to installation and operations.

Oregon
 Currently, Oregon's Department of Environmental Quality (2006) has no air quality requirements for people who make biod-

iesel for their personal use. Producers who make biodiesel for others (e.g., as a co-op, for commercial sale, or for distribution) must file a Notice of Intent to Construct and must include an estimate of air pollution emission with this notice. Producers who will emit over one-half ton of single-criteria pollutants each year must have an Air Contaminant Discharge Permit. Hazardous Air Pollutant emissions exceeding ten tons per year require a Title V Air Operating Permit. Furthermore, air pollution control equipment might be required if your facility emits pollutants above the threshold amounts.

Important Links – Oregon:

Notice of Intent to Construct form: http://www.deq.state.or.us/aq/aqpermit/ACDP/index.htm

DEQ locations: http://www.deq.state.or.us/about/locations.asp.

New Hampshire

Methanol is regulated under the "NH Code of Administrative Rules, Chapter Env-A 1400, Regulated Toxic Air Pollutants." These regulations require demonstrated compliance with emission thresholds of toxic air pollutants. The New Hampshire Department of Environmental Services (NHDES) has evaluated a few small-scale facility plans that produce a daily average of 1,000 gallons (all base-catalyzed transesterification); it has found them to be in compliance without restrictions. All of these plans were reported to have a closed-loop system and store raw materials and waste products in closed containers (NHDES 2006).

Facilities demonstrating compliance still must maintain documentation and present such documentation at the request of NH-DES.

Air permits **are required** if:

Fuel-burning devices do not meet emissions thresholds in Env-A 607.01.
- Example: a boiler rated at 12 million Btu per hour burning B100 (considered "diesel fuel oil" in New Hampshire) requires air permit

Important Links—New Hampshire:
http://www.des.state.nh.us/factsheets/co/inc/16.html

Note: New Hampshire is evaluating how to address stationary sources that burn biodiesel. The Regulated Toxics Air Pollutant rule (Env-A 1400) does not apply to sources burning virgin fuel oil. However, biodiesel blends (anything over B10) are not considered a virgin fuel and, therefore, Env-A 1400 applies to sources burning biodiesel B10 or greater. They are in the process of finding sufficient data to evaluate what regulated toxic air pollutant emissions result from burning biodiesel in stationary sources (e.g., engines, boilers, etc.). A New Hampshire DES representative also mentioned that some producers have proposed burning the glycerin byproduct in boilers, which would also be subject to Env-A 1400. But, again, they have not found sufficient regulated toxic air pollutant emission data from burning glycerin to determine compliance and enforcement (Moore 2007).

Water and Wastewater Requirements

The National Pollutant Discharge Elimination System (NPDES) was established in Section 402 of the Clean Water Act in order to limit pollutant discharges into streams, rivers, and bays. The EPA ultimately has the authority to review applications and permits for "major" discharges, which is a distinction based on the amount and type of discharge. Each state administers the program and issues permits for all point-source discharges to surface waters.

Biodiesel production can result in pollutant discharge through construction-related or operations-related stormwater contamination or through the generation of byproduct wastewater. Managing pollutants without discharging to state waters or a wastewater treatment facility is the best option and results in the least amount of regulatory burden.

Wash water (or wastewater) resulting from biodiesel processing may contain organic residues (esters, fatty acids, soaps, glycerin, and traces of methanol) and inorganic acids and salts produced by the neutralization of residual catalyst with acids (usually hydrochloric acid and sodium chloride). Wash water is high in biochemical oxygen demand (BOD) levels (International Finance Corporation 2007). Wash water that has not had the methanol removed could be categorized as a spent solvent under the Resource Conservation and Recovery Act (RCRA) and as a hazardous waste from a non-specific source (industry and EPA hazardous waste No. F003) and must be managed according to State and Federal regulations (see 40 CFR 261.31 (2006)). Wash water discharged into a municipal sewer, septic system, or a nearby body of water may be allowed or regulated under different standards and regulations that vary from locality to locality.

Some states, such as Virginia, have developed regulations that pertain to the treatment of biodiesel wastewater. If process wastewater will be released to surface water, biodiesel producers will need a Virginia Pollutant Discharge Elimination System (VPDES) permit. If the process wastewater will be applied to land, a Virginia Pollution Abatement Permit is necessary. Additional permitting may also be required if the facility will use an onsite septic system and drain field. Onsite storage tanks for petroleum must be registered to the DEQ Petroleum Program and may be subject to other requirements.

If the facility disturbs wetlands, water bodies or coastal dunes or withdraws large amounts of surface or ground water, producers might require special permits.

Contact your local wastewater treatment plant (often called publicly owned treatment works, or POTWs) to determine if you are required to obtain a discharge permit. Some POTWs are not equipped to treat wastewater containing process chemicals, oils, glycerin, and other contaminants resulting from the biodiesel manufacturing processes. They may therefore require you to treat or neutralize the wastewater before you can discharge it. If you plan to construct a wastewater holding tank or wastewater treatment system, a permit usually is required prior to construction. In New Hampshire, for example, the manufacturer is required to register the discharge under the Underground Injection Control Program if the wastewater is to be disposed of in a septic disposal system (New Hampshire Department of the Environmental Services 2006).

Discharging wastewater directly into any waterways (e.g., streams, rivers, and lakes) requires a National Pollutant Discharge Elimination System permit. The permit typically contains discharge limitations, monitoring, and reporting requirements. Contact your state's EPA Division for details.

Virginia's, Ohio's, and New Hampshire's process water/wastewater regulations are outlined below.

Virginia

Virginia administers the National Pollutant Discharge Elimination System as the Virginia Pollutant Discharge Elimination System (VPDES). The VPDES provides permits for discharges to surface waters for all point sources; however, the US EPA still maintains authority to review "major" dischargers' applications and permits. The Virginia DEQ issues individual and general permits for municipal and industrial facilities as well as for a general class of dischargers. Below is a list of permits that Virginia DEQ has identified as potential required permits resulting from small-scale biodiesel production.

Permits **are required** if:

Process wastewater is recycled on site with the potential to discharge.
    • Virginia Pollution Abatement (VPA) permit required if there is a potential to discharge process wastewater

Process wastewater or sanitary wastewater is discharged to storm sewer or surface waters.
    • Virginia Pollution Abatement (VPA) permit required

Materials and/or processes are exposed to stormwater.
    • VPDES General Permit for Discharges of Stormwater Related to Industrial Activity required

Water permits **are NOT required** if:

Process wastewater is discharged to a sanitary sewer for off-site treatment.
    • Check with your local sewer authority for requirements pertaining to indirect discharges.

Neither materials nor processes are exposed to stormwater.
    • Contingent upon approval of "no exposure certification"

Stormwater, Process Wastewater, and Source Water
    Both the VADEQ and the VA Department of Conservation and Recreation (DCR) regulate the management of pollutants carried by stormwater runoff. "DEQ regulates stormwater discharges

associated with "industrial activities," and the DCR "regulates stormwater discharges from construction sites and from municipal separate storm sewer systems" (VADEQ 2007, 8).

Blow down water from boilers and cooling water systems, run-off from raw materials storage, and wastewater discharged into a municipal sanitary sewer system requires contacting the off-site wastewater treatment facility, which might require some pretreatment before discharge to the sewer system.

Process water and water for equipment operation can be obtained from a public water supply, groundwater wells, or from surface water and might be tied in with siting requirements.

Important Links – Virginia:

Forms for pollutant management activities: http://www.deq.virginia.gov/vpdes/permitfees.html

DEQ regional offices (where forms are submitted):
http://www.deq.virginia.gov/regions/homepage.html

DEQ stormwater regulatory information:
http://www.deq.virginia.gov/vpdes/stormwater.html.

DCR stormwater discharges regulatory information:
http://www.dcr.virginia.gov/soil_&_water/stormwat.shtml

Groundwater withdrawal information:
http://www.deq.virginia.gov/gwpermitting/

Ohio

The Ohio EPA's Division of Surface Water regulates process wastewater discharges. However, local wastewater treatment plants (or publicly owned treatment works (POTW)) must be contacted directly for local requirements and might issue a discharge permit directly. The Ohio EPA allows POTWs to control what enters local sewers but approves pretreatment programs and maintains a list of such programs. Thus, if the POTWs pretreatment plan has not been approved, the Ohio EPA's Division of Surface Water is responsible for issuing the discharge permit (called an indirect discharge permit).

Permits **are required** if:

Process wastewater is discharged directly to any state waters or a conveyance system leading to a waterway.
   • National Pollutant Discharge Elimination System (NPDES) discharge permit required

There is an intention to construct a wastewater treatment system.
   • Permit-to-install required prior to construction

There is an intention to construct, expand, or modify an individual sewage treatment system (rarely applies to small-scale biodiesel production unless you are in an area with no sewers).
   • Permit-to-install required and, if system will have discharge, an NPDES required. Note: only sanitary waste and wastewater can go to an on-site sewage treatment or disposal system and CANNOT include chemicals, oils, or other contaminants (Ohio EPA 2007, 4).

Stormwater, Process Wastewater, and Source Water
   The Ohio EPA regulates the management of pollutants carried off by stormwater. Regulated entities, or facilities falling into any of eleven categories, are required to have a National Pollution

Discharge Elimination System storm water permit. A storm water permit also requires a pollution prevention plan.

Two particular storm water permits apply to biodiesel operations. One is required for construction activities that might disturb one or more acres. A second permit is required if the facility falls into one of the eleven industrial activity categories (see below) in the storm water regulations. Regulated categories are described at http://cfpub1.epa.gov/npdes/stormwater/swcats.cfm and are listed below. The industrial activity-related permits are divided into two further categories – individual and general.

- Category One (i): Facilities with Effluent Limitations
- Category Two (ii): Manufacturing
- Category Three (iii): Mineral, Metal, Oil, and Gas
- Category Four (iv): Hazardous Waste, Treatment, or Disposal Facilities
- Category Five (v): Landfills
- Category Six (vi): Recycling Facilities
- Category Seven (vii): Steam Electric Plants
- Category Eight (viii): Transportation Facilities
- Category Nine (ix): Treatment Works
- Category Ten (x): Construction Activity (see web site for further detail)
- Category Eleven (xi): Light Industrial Activity

If a facility is in one of the eleven categories – except Category Ten – and certifies a condition of no exposure by demonstrating that materials and operations at the site are not exposed to stormwater, the facility might be excluded from industrial stormwater permit requirements.

Industrial biodiesel production falls under the major SIC industrial group 28 (2869) and requires a stormwater permit. Depending on the design and size of your operation, the no-exposure exemption or the SIC classification might apply.

Important Links—Ohio:

Storm Water Program Site:
www.epa.state.oh.us/dsw/storm/index.html

Construction Storm Water Permit: www.epa.state.oh.us/dsw/storm/construction_index.html

Industrial Activity Storm Water Permit: www.epa.state.oh.us/dsw/storm/industrial_index.html

New Hampshire

If wastewater resulting from biodiesel production contains high free fatty acid content and glycerin, has a high biological oxygen demand (BOD), and the manufacturer desires to dispose of it in a septic system, registration is required under NH Underground Injection Control (UIC) Program.

Permits **are required** if:

Wastewater is discharged to surface water.
• National Pollutant Discharge Elimination System permit required (stormwater runoff also regulated under this permit)

Wastewater is disposed of in an existing sanitary collection system.
• Permit to discharge into local system/treatment plant required from municipality
• Industrial Wastewater Indirect Discharge Request (IDR) application must be submitted to the NHDES through

the wastewater treatment facility (see New Hampshire Department of Environmental Services 2006).

Important Links—New Hampshire:

Environmental Permitting, Regulations, and Other Requirements Related to the Manufacture of Biodiesel:
http://www.des.state.nh.us/factsheets/co/co-16.htm

Glycerol and Methanol

Some states have adopted federal regulations found under the Resource Conservation and Recovery Act (RCRA), which describes how waste must be identified, handled, and discarded. Hazardous waste and solid waste are different materials in regulatory terms. A solid waste is a material that no longer has a use, market, or can be recycled legitimately.

According to the EPA's web site, "Hazardous waste is a waste with properties that make it dangerous or potentially harmful to human health or the environment. Hazardous wastes can be liquids, solids, contained gases, or sludges. They can be the byproducts of manufacturing processes or simply discarded commercial products, such as cleaning fluids or pesticides. In regulatory terms, a RCRA hazardous waste is a waste that appears on one of the four hazardous wastes lists (F-list, K-list, P-list, or U-list) or exhibits at least one of four characteristics – ignitability, corrosivity, reactivity, or toxicity." Hazardous waste, regulated under Subtitle C of RCRA, must be sent to a hazardous waste disposal facility.

Byproducts, or "sidestreams," of the biodiesel process include wash water/wastewater, glycerol or glycerin, botched biodiesel batches, unusable waste vegetable oil, and any other creative inventions that handy home brewers and other biodiesel producers make.

This section discusses regulation and permitting in the context of glycerol or glycerin and methanol. (Wash water was discussed in the water and wastewater requirements section.)

Any other solid wastes resulting from the process, including bad biodiesel batches, could be regulated under RCRA if these wastes exhibit one of the four hazardous characteristics listed above (ignitability, corrosivity, reactivity, and/or toxicity) or if the waste is listed specifically as hazardous waste in the EPA's regulations. In order to avoid uncomfortable future visits from your local environmental protection officer, begin an open dialogue with regulators, and keep them apprised of the variables in your operation. Because biodiesel regulations change frequently and vary from location to location, be sure to stay abreast of any developing regulations in your area.

A good small-scale production model that claims to ensure that biodiesel meets ASTM D 6751 fuel quality requirements, avoids toxic waste streams, addresses chemical safety handling, and ensures that biodiesel is produced in an environmentally sustainable manner is featured in William Kemp's book *Biodiesel Basics and Beyond*.

A crude glycerin byproduct typically contains about 50% glycerin with the rest being leftover methanol, catalyst, and so forth. Because of the glycerin's flammability due to the presence of methanol (flash point less than 140 degrees F), this byproduct may be considered hazardous waste.

It must be sent to a permitted hazardous waste disposal facility. Only units regulated and permitted under RCRA as boilers or industrial furnaces may burn hazardous wastes for energy recovery. Furthermore, only facilities with incineration permits from RCRA can incinerate hazardous wastes for disposal.

Virginia is used as a case study to discuss how some states are regulating raw and crude glycerin byproducts.

Glycerol or Glycerin

Biodiesel manufacturing results in a 10% to 20% crude glycerin waste stream. A glycerin byproduct containing residual methanol will put your operation on the radar screen of hazardous waste permitting officials. Although homebrew methanol recovery is on the rise, it is not done by all home brewers. As far as regulations are concerned, integrating methanol recovery into your biodiesel reactor design should decrease the environmental permitting burden. Co-products should have a flash point above 140°F and a pH of less than 12 in order to obtain the "non-hazardous" rating. If the flash point of glycerin is below this specification, either as a raw glycerin byproduct or through methanol recovery, it might be subject to permitting.

Crude glycerin (containing unused catalysts, soaps, methanol, etc.) has little appeal to the market. Some of the creative solutions for the large amounts of glycerin resulting from biodiesel production include composting, combustion in waste oil-burning furnaces, glycerol-fuel cell, industrial lubricants, animal feed, dust suppression, erosion control, and other land applications. Much of this creativity has not moved beyond theory and small demonstration because many states do not allow such applications, especially if methanol is still present in the glycerin byproduct. As mentioned, composting is allowed in some states, but the scale often is regulated. For example, some states only allow composting of glycerin in quantities small enough for a personal garden to handle.

Refined or purified glycerol has many uses in the markets of medicine and pharmaceutical technology, personal care, foods and beverages, and other industries. However, as biodiesel production climbs, so does the amount of glycerin on the market; in fact, so much glycerin has been produced as a result of the biodiesel industry that there is a glut in the market. Author Dave Nilles (2006) has commented, "Saying the domestic crude glycerin market is reaching its saturation point would place one in the running for understatement of the year."

Each biodiesel operation should factor glycerin into the equation and determine how it will be handled and stored before beginning production. A few hundred gallons of biodiesel production per week will produce about 55 gallons of glycerin cocktail. If the methanol is not recovered, this glycerin mix is considered a hazardous substance. Glycerin has a boiling point of 554°F, making purification impossible for the low-budget homebrew operation and, therefore, not recommended.

Check with a local waste motor oil collection facility, industrial composting facility, or landfill to find out who might accept the crude glycerin (once impurities have been removed). Some state environmental departments regulating glycerin disposal will offer advice on glycerin disposal options. Other states might require the development of a glycerin waste disposal plan and check to ensure compliance. Pennsylvania requires characteristic test results prior to accepting glycerol for landfill dumping (Steiman 2007). Robert Miller, the biodiesel processor discussed earlier, states that one auto race track operator uses glycerol for dust control. He dilutes it with water at a ratio of about 5 gallons of glycerol to 6000 gallons of water and applies it before the race; he has reported good dust suppression results. Meanwhile, a local farmer plans to feed it to his cows.

Virginia

Glycerin that **has not** been "de-methylated" – meaning that it still contains methanol – cannot be disposed of through the sewer system, a dumpster, burned, composted, used as fertilizer, or evaporated. Only facilities holding permits under RCRA as boilers, industrial furnaces, or incinerators can process raw glycerin.

If methanol is recovered, records should be kept on site demonstrating that waste streams were evaluated and found to be non-hazardous. Currently, the Virginia DEQ has no reference materials suggesting that it would be beneficial to apply a glycerin byproduct as a soil amendment or compost additive (VADEQ 2007, 15). Waste generators are allowed to submit information to the VADEQ, including "a full regulatory analysis of any exclusion applicability and adequate demonstration that the material in question is an effective substitute for another material typically applied to land and in commercial use" (VADEQ 2007, 15). The Virginia Department of Agriculture and Consumer Services' also would need to approve such applications and would require a description of scenarios for acceptable soil amendment or fertilizer use and application rates.

Methanol Recovery

Byproducts exhibiting hazardous waste characteristics are not solid wastes when reclaimed under the provisions of 40 CFR 261.2 (see table below). According to VADEQ, recovering and reusing methanol "is considered a reclamation process" and is "excluded from most solid or hazardous waste regulations" (2007, 15).

Secondary materials from the biodiesel process are regulated by RCRA if they are determined to be solid wastes when they are recycled. How do you know if something will be considered a solid waste when you recycle it? According to RCRA Subtitle C, materials that are "accumulated speculatively" before being recycled are considered solid wastes (261.2(c)(4)).

If this leaves you feeling a bit unclear and scratching your head, you are not alone. The EPA actually created a speculative accumulation provision to reduce the risk of people/companies accumulating a lot of hazardous secondary materials to be recycled. Essentially, the EPA says that you cannot over-accumulate such materials to the point of creating an environmental hazard. So, if you have piles of glycerol that you are saving for a rainy, soap-making day, you might want to reconsider.

	Use constituting disposal (§261.2(c)(1))	Energy recovery/fuel (§261.2(c)(2))	Reclamation (§261.2(c)(3)) (except as provided in 261.4(a)(17) for mineral processing secondary materials)	Speculative accumulation (§261.2(c)(4))
	1	2	3	4
Spent Materials	(*)	(*)	(*)	(*)
Sludges (listed in 40 CFR Part 261.31 or 261.32)	(*)	(*)	(*)	(*)
Sludges exhibiting a characteristic of hazardous waste	(*)	(*)	___	(*)
By-products (listed in 40 CFR 261.31 or 261.32)	(*)	(*)	(*)	(*)
By-products exhibiting a characteristic of hazardous waste	(*)	(*)	___	(*)
Commercial chemical products listed in 40 CFR 261.33	(*)	(*)		___
Scrap metal other than excluded scrap metal (see 261.1(c)(9))	(*)	(*)	(*)	(*)

Note: The terms "spent materials," "sludges," "by-products," and "scrap metal" and "processed scrap metal" are defined in §261.1.

Table Indicating When Certain Materials Are Regulated As Solid Wastes or Are Exempt from Regulation When Reclaimed. *Adapted from US Government Printing Office 2005.*

Specifically, the EPA "subjects persons who 'accumulate speculatively' (i.e., persons who fail to recycle a sufficient percentage of a recyclable material during the calendar year or fail to demonstrate that a feasible means of recycling exists) to immediate regulation as hazardous waste generators or storage facilities. The speculative accumulation provision generally applies to secondary materials that are not solid wastes when recycled. In other words, certain secondary materials that are otherwise excluded from the definition of solid waste become regulated as solid and hazardous waste if accumulated speculatively" (U.S. Environmental Protection Agency 1995).

If a facility wishes to avoid exceeding speculative accumulation timeframes, it must reclaim at least 75% of the materials (e.g., methanol) within the calendar year following the generation event. Therefore, a biodiesel producer must reclaim 75% of the methanol used within a year of producing the raw glycerin byproduct. After this rule is exceeded, the conditional exclusion as a recyclable material is lost, and solid and/or hazardous waste management requirements are applicable (VADEQ 2007, 15).

Records must be maintained documenting legitimate recovery for reuse/resale if a facility wishes to claim a regulatory exemption for off-site reclamation of glycerin or methanol or use by a glycerin refinery. Claims for a conditional exemption are to be backed up by "analytical results, contracts with reclamation facilities, shipping documents" (VADEQ 2007, 16).

For additional information, you might wish to view the EPA Training Module: Introduction to Hazardous Waste.
http://www.epa.gov/epaoswer/hotline/training/hwid05.pdf

On-site Storage and Transport of Production Materials

Federal and state laws have requirements that apply to aboveground and underground storage tanks with a certain volumetric capacity. Local building officials, fire marshals, and Departments of Environmental Quality or Natural Resources should be consulted prior to construction for the required permits, installation requirements, and local codes and ordinances.

Federal law requires individuals with an aboveground combined storage capacity of over 1,320 gallons or an underground storage capacity of over 40,000 gallons of any type of oil (e.g., petroleum, mineral, vegetable) to prepare a Spill Prevention, Control, and Countermeasures (SPCC) plan. This plan should outline the measures that you would take to control and respond to an oil-related spill on site. The EPA enforces the requirements for this plan and inspects the SPCC plan-holder's facilities (US Environmental Protection Agency 2006). Secondary containment usually is required for oil/oil product storage units and should be considered during the early planning stages of your operation. Contact your state's

EPA for more information on SPCC plans and biodiesel storage requirements.

Some states do not have rules regulating the storage of B100 either above or below ground because biodiesel is not considered to be a petroleum product, hazardous waste, or a wastewater. However, once diesel fuel is blended with B100, the petroleum underground or aboveground storage tank rules apply. Your Departments of Environmental Quality or Natural Resources would have jurisdiction over this matter. As a result, many biodiesel cooperatives have chosen to store B100 for members, who use it in biodiesel blends off site.

Similarly, the transport of B100 also is facilitated by its non-hazardous rating. However, this does not exclude fuel transporters from needing to obtain licenses, filing a bond, and so forth.

Methanol storage might be limited to two 5-gallon containers, similar to gasoline storage cans, in some jurisdictions (Andreas 2005).

In some cases, used restaurant cooking oil is regulated. For example, in Michigan, it is considered a liquid industrial waste if someone other than the "generator" (person who used it) collects and transports it. As a result, both the collector of the oil and its generator must be licensed (Leidel Energy Services 2006).

Since the EPA Spill Prevention Control and Countermeasure (SPCC) requirements apply in each state, there is a brief discussion of the regulations below. An overview of Virginia's, Ohio's, and Pennsylvania's transport and storage regulations for biodiesel are shown below.

EPA Spill Prevention Control and Countermeasure (SPCC) Requirements

The U.S. Environmental Protection Agency (EPA) Oil Pollution Prevention Regulation (40 CFR 112) applies to non-transportation-related facilities that store oil or petroleum products in quantities above threshold amounts. For those facilities that are regulated, the primary requirement is that they create and implement a Spill Prevention, Control, and Countermeasure (SPCC) Plan that will prevent any oil from entering US waters. According to the EPA, this regulation "applies to any onshore or offshore facility engaged in drilling, producing, gathering, storing, processing, refining, transferring, distributing, using, or consuming oil and oil products, providing that all three of the following conditions are met:

1. The facility is non-transportation-related;
2. The aggregate aboveground storage capacity is greater than 1,320 gallons, with a de minimus container capacity of 55 gallons, or the total underground storage capacity is greater than 42,000 gallons; and
3. Due to its location, oil discharged at the facility could reasonably be expected to reach waters of the United States or adjoining shorelines" (U.S. Environmental Protection Agency 2006).

An SPCC plan addresses operating procedures to prevent oil discharges, control measures to prevent any oil discharge from entering navigable waters, and countermeasures to contain, clean up, and mitigate the impacts of a discharge to navigable waterways. The SPCC plan must be approved by a Professional Engineer if the conditions above are met. For more information, refer to the EPA's online fact sheet: http://www.epa.gov/reg3hwmd/oil/spcc/index.htm.

Virginia

Virginia DEQ's Facility and Aboveground Storage Tank Regulation (9 VAC 25-91-10 *et seq*) contains requirements for aboveground storage tanks that have a volumetric capacity over 660 gallons and contain "a liquid hydrocarbon product or mixtures of hydrocarbon production with other products" (VADEQ 2007, 9).

Virginia DEQ's Underground Storage Tanks: Technical Standards and Corrective Action Requirements (9 VAC 25-580-10 *et seq*) contains requirements for underground storage tanks that have a volumetric capacity over 110 gallons and contain "regulated substances, including petroleum, biodiesel blends, or hazardous substances" (VADEQ 2007, 9).

Important Links – Virginia:

DEQ Storage Tank Programs:
http://www.deq.virginia.gov/tanks/stortnks.html

EPA SPCC Plan Program:
http://www.epa.gov/reg3hwmd/oil/spcc/index.htm

Ohio

Those handling oil or oil products (including vegetable oils) could be subject to the SPCC regulations (40 CFR Part 112).

Facilities are subject to SPCC regulations if:

- Aboveground storage capacity exceeds 1,320 gallons; or
- Underground storage capacity exceeds 42,000 gallons ("excluding tanks regulated under 40 CFR Parts 280 or 281") (Ohio Office of Compliance Assistance and Pollution Prevention 2007)

The State Fire Marshal's Bureau of Underground Storage Tank Regulations also apply.

For more information, refer to the Ohio's EPA fact sheet titled, "Understanding the Spill Prevention, Control and Countermeasure (SPCC) Requirements."

http://www.epa.state.oh.us/ocapp/sb/publications/spcc.pdf.

Pennsylvania

Pennsylvania regulates aboveground hazardous substance storage if capacity is greater than 250 gallons; it regulates underground storage if capacity is greater than 110 gallons. On-farm storage of biodiesel blends is not regulated if capacity is less than 1,100 gallons. A Preparedness, Prevention, and Contingency Plan typically is required if the hazardous materials are stored in quantities greater than 55 gallons without secondary containment. Although unprocessed oil storage is not regulated, waste oils fall under Pennsylvania Department of Environmental Protection Waste Management storage regulations (Steiman 2007). The Pennsylvania Department of Labor and Industry and/or the local fire marshal (in Alleghany and Philadelphia, storage regulation is under fire marshal jurisdiction) regulates the storage of B60 or lower that is intended for vehicular use but does *not* regulate storage of biodiesel used for heating or power generation.

Codes, Standards, and Fire Marshals

Waste grease, methanol, and finished biodiesel are combustible liquids (New Hampshire Department of Environmental Services 2006). Fire departments, building officials, code enforcement officers, and zoning administrators should be consulted for compliance with all state and local codes.

Building Officials and Code Administrators (BOCA), International Organization for Standardization (ISO), Uniform Building Code (UBC), Uniform Fire Code (UFC), or other codes used by the agency that serves your location are some of the codes that you might encounter. However, codes and standards vary by locale. Thus, codes that apply to another group in another location might not apply to your operations. Depending on where you are located, what you store on site, and how you store it will determine which codes apply. If there are no biodiesel production-related codes or standards for your area, find out what precedents have been set in surrounding areas. Then, provide that information to fire or building officials since they prefer to work from precedent, but do not be surprised if they decide to act differently than officials in neighboring areas. It also is a good idea to pay a visit to the group that provided you with the codes and standards for your area to ensure a mutual understanding of your intentions with your local officials.

Compliance with current fire and building codes sometimes requires the construction of a "high hazard building" due to the flammability of the methanol that often is used in biodiesel processing. The "hazardous" designation is based on the potential storage capacity of any involved vessels in which hazardous material is stored rather than the actual amounts used. Constructing a compliant "high hazard area" can cost a few hundred thousand dollars and can be cost prohibitive to small operators. Because regulations vary considerably by location, it is imperative that biodiesel producers understand the regulations associated with their specific locations.

Local fire marshals have considerable safety expertise and often have a good deal of discretion, so it is worth developing a good relationship with the fire marshal and executing his/her suggestions. Also note that fire marshals might be learning about biodiesel just as you are; keep an open dialogue so that you can inform one another mutually. Biodiesel is considered a Class IIIB combustible liquid according to the International Fire Prevention Code, which is not a flammable substance and, therefore, reduces the amount of regulatory requirements. Fire officials may want to confirm this using a "closed cup flash point test."

Fire officials often think of biodiesel in terms of its manufacturing stages. For example, biodiesel production starts as a feedstock (safe with a high flash point). Next comes the methanol catalyst (flammable liquid, low flash point, poisonous, wide flammability range, burns with no visible flame) and sodium hydroxide (reactive, caustic, high heat of solution). The combination of these substances creates methoxide, which is flammable, poisonous, and reactive with water (International Code Council 2007). Therefore, these last two stages mentioned might be a cause for concern for code officials, fire marshals, and regulators. Understand that even if biodiesel production is "safe and easy" in comparison to other manufacturing processes, it still presents hazards, and you can burn your house or farm down and injure yourself and others.

Because biodiesel production can be a fire hazard, the question of sprinklers sometimes arises. The maximum allowable quantity of a Class IIIB liquid inside one control area of a non-sprinkled building is 13,200 gallons.

If you are not on zoned agricultural land and you decide to place a pumping station on location, get a permit. Beginning the process without doing so might irk your fire marshal and local building inspector. Biodiesel expert Lyle Estill wrote about this issue in his widely read Energy Blog. Piedmont Biofuels decided to house an existing pumping station inside a straw bale structure after a cold winter snap. This did not go over well with the state authorities. After arguing that there were no code references for storing B100 in a straw bale structure and appealing the case to the State Fire Marshal on the basis that B100 and straw bale do not burn very well, Piedmont Biofuels prevailed. However, when the Inspections Department came for the assessment, Piedmont Biofuels was ordered to drywall the straw bale structure and, on the second inspection, to fix the placement of a battery box (Estill 2006). Piedmont Biofuels finally received the coveted Certificate of Occupancy but not without much time and energy expended on a small straw bale dispensing station.

The fundamental message appears to be to contact the proper authorities prior to any construction, storage, or operation activities. This might just help to keep officials from shutting your operation down before you even get started.

Emergency Planning and Community Right-to-Know Act (EPCRA) Requirements

EPCRA, or Title III of the Superfund Amendments and Reauthorization Act (SARA), aims to promote planning for chemical emergencies and to provide the public with information on chemicals that are stored, used, and released in their community. EPCRA has four major components, including emergency planning notification (section 302), emergency release notification (section 304), hazardous chemical inventory reporting (sections 311 and 312), and toxic chemical release inventory (section 313). Basically, the act requires local emergency response authorities to be informed of any hazardous chemicals that are stored on site as well as discharges or environmental releases of chemicals beyond threshold amounts. Section 313 of EPCRA and some state laws require reporting of use, recycling, energy recovery, treatment, disposal, emissions, and releases of chemicals listed in the Toxic Release Inventories (VADEQ 2007, 12).

Each section of EPCRA lists different criteria used in determining reporting requirements to a state's emergency response commission (SERC) and/or a local emergency planning committee (LEPC). Section 302 deals with extremely hazardous substances; therefore, it generally does not apply to biodiesel production. However, a breakdown of the remaining EPCRA sections is shown below:

Sections 311 and 312 of EPCRA require the reporting of on-site hazardous chemical storage to local emergency authorities when OSHA Hazard Communication Standards apply (requiring an MSDS for a chemical) as when hazardous chemicals and/or extremely hazardous substances are used, produced, or stored (U.S. Environmental Protection Agency 2000).

According to Sections 311-312, methanol, sodium and potassium hydroxide, and glycerin are hazardous chemicals with threshold quantities of 10,000 pounds for each chemical (not for the combination of chemicals).

Section 304 of EPCRA requires reporting a release or discharge of a chemical if the reportable quantity is exceeded and crosses your property's boundary. Chemicals subject to reporting requirements include extremely hazardous substances (40 CFR Part 355; Appendix A and B), CERCLA hazardous substances (40 CFR PART 302; Table 302.4), and oil (VADEQ 2007, 10).

Under Section 304, methanol's reportable quality is 5,000 pounds.

Section 313 of EPCRA requires states to establish a Toxics Release Inventory of "toxic chemical use, recycling and energy recovery, treatment and disposal, and releases and emissions from certain types of facilities" (VADEQ 2007, 10).

If all of the following conditions are met, a report must be filed if the facility:

- Has 10+ full-time employees
- Is included in a list of Standard Industrial Classification (SIC) codes
- Manufactures, processes, or uses listed toxic chemicals in quantities above the threshold. Most chemical threshold quantities are 10,000 pounds per year for use and 25,000

pounds per year for manufacturing and processing. (VADEQ 2007, 10)

Methanol is subject to TRI reporting if threshold quantities are exceeded.

The Georgia Environmental Compliance Assistance Program has developed some helpful flowcharts on the EPCRA provisions described above to help facilities determine reporting requirements.
http://www.gecap.org/

Occupational Health and Safety

Those involved with biodiesel also need to consider some occupation health and safety issues, such as process safety, chemical hazards, fire and explosions, and other occupational hazards (International Finance Corporation 2007). Some of these were discussed in the chapter on safety. How you manage these potential hazards in the workplace depends on your operation and whether or not employees are involved.

Your local Occupation Health and Safety Administration (OSHA) program, which usually falls within your state's Department of Labor and Industry, should be able to assist you with any OSHA compliance requirements. Additionally, obtain, read, and keep material safety data sheets on site and in every transport vehicle for the following chemicals: methanol, ethanol, sodium hydroxide or potassium hydroxide, sulfuric acid, glycerin, and biodiesel. There is a sample biodiesel MSDS located at www.biodiesel.org.

Spills

The EPA regulates spills of more than 40 gallons of oil. Report spills to the National Response Center (1-800-424-8802 or http://www.nrc.uscg.mil/nrchp.html). Spills also are covered under the

US EPA Oil Pollution Prevention regulations, which were discussed earlier in this chapter.

Requirements for Biodiesel Resellers

EPA Registration

The first issue that needs to be addressed when considering producing biodiesel for commercial sale in the US is registration as a biodiesel fuel producer with the EPA. In the US, the EPA governs fuel and fuel additive registration, and anyone selling biodiesel first must be registered with that agency by providing the EPA with a laboratory analysis indicating the fuel's compliance with ASTM D 6751.

Furthermore, fuels and fuel additives in the US must be registered with the US EPA and must meet the health regulations found in 40 CFR Part [79]. The EPA allows producers to access this information from a group since it is very costly. The most popular option utilized is National Biodiesel Board (NBB) membership, which is discussed below.

Manufacturers can register their fuel with the EPA by completing a registration form and supplying additional information. The registration form (Form 3520-12) is available from the EPA's web site at www.epa.gov/otaq/regs/fuels/ffarsfrms.htm. Following all required documentation approval (laboratory analysis indicating full compliance with ASTM D 6751 AND health effects data), the EPA usually will grant registration within one month.

National Biodiesel Board Registration

In order to register fuel with the EPA for on-road use, producers must have access to health affects testing data on the fuel being registered. This requirement could be difficult on individuals and small-scale producers. However, the EPA allows producers to access this information from a group, such as the National Biodiesel Board, which finished its testing in 1997 and has its data on file with the EPA. All of the NBB's biodiesel processor and small-scale producer members can access this information for free (National Biodiesel Board 2006b). A non-voting, small-scale producer member is one that makes fewer than 250,000 gallons of biodiesel per annum. To become a member of the NBB, the producer must verify that "it will process only biodiesel that is registered with the EPA pursuant to the Clean Air Act regulations found at 40 C.F.R. Part 79" (National Biodiesel Board 2006b). Information about NBB membership can be accessed through the company's web site at http://www.biodiesel. org/members/info/.

Biodiesel Producer, Importer, and Blender Registration

EPA

Commercial biodiesel producers and importers need to follow the guidelines set forth by the Federal Renewable Fuels Standard (FRFS), which went into effect on September 1, 2007. According to this standard, all biodiesel producers, importers, exporters, and refiners must register with the US EPA. Registering companies will be given a four-digit company identification number; however, those companies that previously registered under the Motor Vehicle Diesel Fuel Sulfur Program do not need to re-register. The new FRFS requires companies to assign unique renewable identification numbers (RINs) to each batch of biodiesel that is produced, refined, imported, or exported. The RINs are 38-character numeric codes that can be traded like carbon credits. For more information, contact the US EPA.

IRS

Before making or importing biodiesel for commercial sale, US biodiesel producers or importers must register with and have their applications approved by the Internal Revenue Service (IRS). Producers/importers will use the updated IRS Form 637 and will register as a producer, an importer, or both depending on what they make – "biodiesel" (Activity Letter "NB") or "agri-biodiesel" (Activity Letter "AB"). NB is biodiesel made from non-virgin feedstock while AB is biodiesel made from virgin feedstock.

Blenders of biodiesel must receive a 637 M designation from the IRS by applying for it on Form 637 (use Activity Letter "M") and must obtain approval prior to starting blending activities.

The updated Form 637 is available on the Forms and Publications page of the IRS web site, which is located at www.irs.gov.

Tax Credits for Biodiesel

The American Jobs Creation Act of 2004 outlined fuel tax provisions, which the IRS is responsible for implementing. These provisions include a tax credit, referred to as the Credit for Biodiesel Mixtures, that applies to biodiesel that is sold or used in business. This credit is 50 cents per gallon of biodiesel (NB) or $1.00 per gallon of agri-biodiesel (AB) that you produce and either sell for fuel or use as a fuel in your own trade or business (Hagen 2005). According to this Act, a biodiesel mixture credit is forbidden unless the blender obtains a certificate from the biodiesel producer that identifies the product as EPA registered, ASTM D 6751 certified biodiesel or agri-biodiesel.

The biodiesel that qualifies for this tax credit consists of "monoalkyl esters of long chain fatty acids" and meets both the EPA registration requirements for fuels and fuel additives and the requirements of ASTM D 6751.

The first step is to follow the procedures for IRS registration as a biodiesel producer, importer, and/or blender. Then, a Certificate for

Biodiesel must be obtained prior to making a claim to the IRS for credits and/or refunds. This certificate can be obtained through the biodiesel producer or the biodiesel reseller. A copy of the certificate must accompany the claim.

The National Biodiesel Board maintains tax incentive information on its site, which includes guidance on all of the various distribution and inventory storage scenarios. Visit this site for clarification and sample scenarios.

Eligible entities that file claims with the IRS for credits and/or payments must follow the procedures outlined in Notice 2005-04 and Notice 2004-62. The following steps are mandatory:

1. "Blenders must use Form 720, Quarterly Federal Excise Tax Form; and Form 8849 to claim their credit(s) and payment for the excess sum of their credit(s) above their excise tax liability" (National Biodiesel Board n.d.).

2. "Blenders must use Form 4136, Credit for Federal Tax Paid on Fuels; or Form 8846 Biodiesel Fuels Credit when claiming an income tax credit" (National Biodiesel Board n.d.).

Additionally, the Energy Policy Act of 2005, commonly referred to as H.R.6, created a credit for small agri-biodiesel producers that is $0.10/gallon on the first 15 million gallons. This credit can be "passed through to farmer owners of a cooperative and is allowed to be offset against the alternative minimum tax" (Renewable Fuels Association 2005). The credit is set to expire in 2008.

Fuel Dealer License

In most cases, each fuel supplier first must be registered in the state in which biodiesel is being sold, and suppliers must be properly licensed, bonded, and insured. Although many states do not have a specific biodiesel fuel dealer's license, biodiesel that is sold to others for use in on-highway motor vehicles may require a diesel fuel dealer's license. Applications typically are filed with the Commissioner of Motor Vehicles at the Department of Motor Vehicles.

Motor Fuels Tax Bond

Some states require anyone wishing to distribute fuel to obtain a certificate of deposit that would be cashed by the state's Department of Revenue in the event of a default on tax payments. This certificate of deposit is often referred to as a motor fuels tax bond.

Business License

Cities require any business opening within city limits to obtain a business license. Contact your local Commissioner of the Revenue. Note that applications often must be approved by zoning boards. Some farm and non-profit entities might qualify as tax-exempt.

Dye and Taxes

Fuels that are used for non-road purposes, such as kerosene and diesel fuels, generally are not subject to taxes and are required to be dyed with Solvent Red 164, also known as Oil Red B. This helps to distinguish them from on-road fuels. Unlike non-road kerosene and diesel fuel, 100% bio-fuel "liquid" does not need to be dyed for off-road, untaxed use according to state and federal legislation (see 26 USC 4041(b)(1)(a)).

Those using fuel with untaxed off-road diesel should follow dyeing requirements, which are 3.9 pounds of solid dye per 1,000 barrels of oil. Ensure that biodiesel blends meet this requirement because there are severe fines for not meeting the minimum dye requirement.

Insurance

Small-scale producers might find that insurance is unavailable or that the costs are prohibitive. In fact, one biodiesel manufacturer explained that its insurance company specifically asked that the manufacturer avoid referring other small-scale producers to the insurer (Estill 2007a). This hesitation to insure biodiesel production can impair the biodiesel business and complicate possible investment or funding opportunities and offers one more reason for staying on-farm or teaming up with a university or a municipality.

Not all of the regulations and requirements discussed above apply to each and every small-scale producer. The information is meant to familiarize potential producers with what they may be up against. As described in Matt Steiman's case study (Dickinson College), an opportunity exists for small-scale producers to work with regulators to develop guidelines that will affect the future of home brewing in their states. A few things to keep in mind as you begin a conversation with your local and state government regulators is that they know what they are doing and that they have the power to prevent you from undertaking small-scale production. This is not the intent of most; however, many regulators are somewhat unfamiliar with biodiesel and the differences between it and diesel fuel. Education is key! Come prepared to any meeting and teach them about biodiesel and your production process, and demonstrate the positive impact that you will have on the community. Document all aspects of your biodiesel production initiative. This will help the regulators to assist you in your efforts to operate safely and legally. Invite them to your biodiesel facility to make some fuel. If you make it fun (and safe), they are likely to become fans and want to help you succeed.

Case Study: Yellow Biodiesel, Williamsburg, Massachusetts

In 1998, Tom Leue bought a diesel tractor and read Joshua Tickell's *From the Fryer to the Fuel Tank*. He has been a fan of biodiesel ever since. Through his company, Yellow Biodiesel, Tom started out as a small-scale producer in western Massachusetts. However, a terrible fire changed things. Like many small-scale producers, he operated on a shoestring budget, purchased the equipment that he could afford – such as a stirrer that was not explosion-proof rather than an explosion-proof one that cost ten times more – and started making biodiesel on his own. Having worked as a chemist, he was undaunted by the chemistry of biodiesel.

However, a fire "burned the factory to the ground," and Tom "barely got out alive." Rather than trying to rebuild, Tom decided to change Yellow Biodiesel's focus to one of distribution rather than production. By focusing on distribution, he avoids many of the regulatory hurdles related to biodiesel and finds that the tax structure is easy to comply with. In fact, he suggests that those interested in small-scale production become distributors first. He says, "The road to production of biodiesel should start with distribution." By selling other producers' biodiesel first, potential small-scale producers can learn the ropes, build an infrastructure, and educate the market. He promotes his packaged biodiesel distribution system, which costs about $75 to set up in every participating store. This helps the small-scale producer to avoid the large-scale costs associated with setting up underground storage tanks and pumps at each service station.

Now, Tom focuses on buying and selling biodiesel. The feedstock varies depending on the batch with some containing mixtures of virgin oil, waste oil, and tallow. His Yellow Brand Biodiesel is sold in yellow plastic, five-gallon containers that can be cleaned out and returned to participating retailers. It is available in many stores in the New England area, and Tom also has developed a franchise system whereby others can distribute biodiesel using his system. Deliveries are made using an old diesel ambulance with shelves, and empty containers are taken back for cleaning and refilling. The ambulance's red lights have been replaced with yellow ones, and Tom says that he takes it to festivals and fairs.

Using his experience, Tom also has helped four other biodiesel groups get off the ground. He suggests that those interested in safely making biodiesel should not go it alone. Collecting waste oil, making biodiesel, and marketing the biodiesel is a big project that "is too much for one individual to do."

Chapter 6. State-by-State Biodiesel Incentives & Regulations

This section highlights some of the state incentives and regulations that relate to biodiesel research, production, storage, distribution, fuel sales, and fuel use.

For additional and updated information on state laws on biodiesel production and for additional contacts, refer to the US DOE's Alternative Fuels Data Center's web site at http://www.eere.energy.gov/afdc/progs/all_state_summary.cgi?afdc/1.

If you want to learn more about your state's alternative fuel incentives or regulations, contact your state representative for clarification on what your state includes in its definition of alternative fuels. For some states, "alternative fuels" refers to such fuels as compressed natural gas, liquefied natural gas, and liquefied petroleum gas (propane). Other states include biodiesel, ethanol, and other fuels in their definition of "alternative fuels." Assume that all biodiesel incentives and regulations refer to biodiesel that meets the ASTM specification for biodiesel.

Alabama

Incentives: None known
Regulations: None known

Contacts:

Mark Bentley, Executive Director, Alabama Clean Fuels Coalition (Not Yet Designated), (205) 402-2755, mark@alabamacleanfuels.org

Keith Fordham, Clean Cities Coordinator, South Alabama Regional Planning Commission/Clean Cities of Lower Alabama (Not Yet Designated), (251) 232-0152 or (251) 208-5893, keithvf@comcast.net

Steven Richardson, Project Manager, U.S. Department of Energy, National Energy Technology Laboratory, (304) 285-4185, steven.richardson@netl.doe.gov

Voris Williams, Air Quality Coordinator, Regional Planning Commission of Greater Birmingham, (205) 264-8448, vwilliams@rpcgb.org

Kathy Hornsby, Program Manager, Alabama State Energy Office, Alabama Department of Economic and Community Affairs, (334) 242-5284, kathy.hornsby@adeca.alabama.gov

Greg Roberts, Director, Business Development, Mobile Gas Service, (251) 450-4742, groberts@mobile-gas.com

Marilyn Franklin/Loretta Cook, License and Fees Secretary, Liquefied Petroleum Gas Board, (334) 242-5649, lcook@lpgb.state.al.us

Dale Aspy, Environmental Engineer, Region 4 Air Planning Branch, U.S. Environmental Protection Agency, (404) 562-9041, aspy.dale@epa.gov

Wes Allen, Transportation Specialist, Southeast Region, U.S. General Services Administration, (404) 331-3052, james.allen@gsa.gov

Alaska

Incentives: None known

Regulations: The state's Department of Transportation must consider using alternative fuels whenever practical.

Contacts:

Ernie Oakes, Project Manager, U.S. Department of Energy, Golden Field Office, (303) 275-4817, ernie.oakes@go.doe.gov

Barbara Shepherd, Environmental Program Specialist, State of Alaska Environmental Conservation - Air Nonpoint and Mobile Sources, (907) 465-5176, barbara.shepherd@alaska.gov

Cary Bolling, Energy Specialist, Alaska State Energy Office, Alaska Housing Finance Corporation, (907) 338-8164, cbolling@ahfc.state.ak.us

Steve Morris, Environmental Quality Program Manager, Municipality of Anchorage, (907) 343-6976, morrisss@muni.org

John Huzey, Fleet and Facilities Manager, Municipality of Anchorage, (907) 343-8312, huzeyjm@muni.org

Craig Lyon, Transportation Planning Manager/AMATS Coordinator, Anchorage Metropolitan Area Transportation Solutions (AMATS), (907) 343-7991, lyonch@muni.org

Bonnie Richard, Alaska Fleet Manager, U.S. General Services Administration, Alaska Fleet Management Center, (907) 271-3940, bonnie.richard@gsa.gov

Arizona

Incentives: None known

Regulations: State agencies and political subdivisions that operate an alternative fueling station must allow all state agencies and political subdivisions to refuel their vehicles at that station. Certain counties that send out requests for proposals for work involving heavy-duty diesel equipment should encourage clean diesel technologies and fuels, biodiesel, or other cleaner alternatives. All biodiesel sold in Arizona is required to meet the ASTM D 6751 specifications while all biodiesel blends must meet the ASTM D975 specification. Biodiesel blenders and dispensers are subject to labeling and reporting requirements. Local governments in particular areas with large populations must develop and put into effect a vehicle fleet plan that supports and enhances the use of cleaner fuels, including biodiesel.

Contacts:

Bill Sheaffer, Executive Director, Valley of the Sun Clean Cities Coalition, Inc., (480) 314-0360, bill@cleanairaz.org

Corey Woods, Clean Cities Coordinator, Valley of the Sun Clean Cities Coalition, (602) 258-7505 x17, corey@cleanairaz.org

Colleen Crowninshield, Clean Cities Manager, Tucson Clean Cities Coalition, (520) 792-1093 x426, ccrowninshield@pagnet.org

Mike Bednarz, Project Manager, U.S. Department of Energy, National Energy Technology Laboratory, (412) 386-4862, michael.bednarz@netl.doe.gov

Arizona Department of Transportation, Motor Vehicle Division Customer Call Center, (602) 255-0072 (Phoenix), (520) 629-9808 (Tucson)

Collette Craig, AFV Contact, U.S. General Services Administration, Region 9, (928) 524-1465, collette.craig@gsa.gov

Arkansas

Incentives: Alternative fuel grants are available to fuel producers, feedstock processors, and distributors through July 1, 2009. Biodiesel suppliers are entitled to a per-gallon tax refund. Additionally, biodiesel suppliers can claim an income tax credit for a portion of their facilities and equipment costs.

Regulations: The state of Arkansas has set an annual sales goal of 50-plus million gallons of alternative fuels. All state-owned or leased diesel-powered equipment and vehicles must contain a minimum of 2% biofuels; certain exceptions apply. Alternative fuels excise taxes have been established.

Contacts:

John R. Hoffpauer, Clean Cities Coordinator, Central Arkansas Clean Cities Coalition, (501) 372-3300, john.hoffpauer@metroplan.org

Randy Thurman, Executive Director/Clean Cities Coordinator, Arkansas Environmental Federation/Central Arkansas Clean Cities Coalition, (501) 374-0263, rthurman@environmentark.org

Neil Kirschner, Project Manager, U.S. Department of Energy, National Energy Technology Laboratory, (412) 386-5793, neil.kirschner@netl.doe.gov

Chris Benson, Director, Arkansas Energy Office, (800) 558-2633 or (501) 682-0865, cbenson@1800arkansas.com

Mike Porta, Assistant Division Chief, Arkansas Department of Environmental Quality, (501) 682-0730, porta@adeq.state.ar.us

Angela Marsh, Loan Program, Arkansas Department of Environmental Quality, (501) 682-0709 or (888) 233-0326, marsha@adeq.state.ar.us

Sandra Rennie, Mobile Source Team Leader, Region 6, U.S. Environmental Protection Agency, (214) 665-7367, rennie.sandra@epa.gov

Gordon Lancaster, Transportation Operations Specialist, U.S. General Services Administration, (303) 236-7599, gordon.lancaster@gsa.gov

California

Incentives: The Air Resources Board's Innovative Clean Air Technologies Program provides co-funding for projects that focus on cleaner fuels and includes projects related to petroleum diesel fuel and diesel engines.

Regulations: The Department of Food and Agriculture's Division of Measurement Standards has developed specifications for biodiesel that will be used in a vehicle. Biodiesel blends or biodiesel that will be used for blending must meet ASTM specifications. Biodiesel blends must be labeled to indicate the percentage of biodiesel contained within. Utilities, solid waste collection vehicles, and public agencies are allowed to use biodiesel or biodiesel blends up to B20. The state also has created a mandate to meet 20% of its energy needs with biofuels by 2010, 40% by 2020, and 75% by 2050. The California Energy Commission and Air Resources Board will work together as part of the Bioenergy Interagency Working Group to develop a Bioenergy Action Plan. Furthermore, the City of San Francisco will require all of the diesel vehicles that are used by the city's public agencies to use a minimum of B20 by December 31, 2007; from that point on, all of the city's public agencies should work to use increasingly higher biodiesel blends up to B100.

Contacts:

Bret Banks, Clean Cities Coordinator, Antelope Valley Clean Cities Coalition, (661) 723-8070, bbanks@avaqmd.ca.gov

Melissa Guise, Clean Cities Coordinator, Central Coast Clean Cities Coalition, (805) 781-4667, mguise@co.slo.ca.us

Linda Urata, Clean Cities Coordinator, San Joaquin Valley Clean Cities Coalition, (661) 835-8665, info@projectcleanair.org

Heloise Froelich, Clean Cities Coordinator, Los Angeles Clean Cities Coalition, (213) 978-0854, heloise.froelich@lacity.org

Chris Ferrara, Clean Cities Coordinator, East Bay Clean Cities Coalition, (925) 674-6533, caf3@pge.com

Rick Ruvolo, Clean Cities Coordinator, San Francisco Clean Cities Coalition, (415) 753-1136, rrsf@aol.com

Jill Egbert, Clean Cities Coordinator, Greater Sacramento Clean Cities Coalition, (530) 757-5235, jme3@pge.com

JoAnn Armenta, Clean Cities Coordinator, Southern California Association of Governments (SCAG) Clean Cities Coalition, (909) 396-5757, joann@the-partnership.org

Vivian Ozuna, Clean Cities Coordinator, Long Beach Clean Cities Coalition, (562) 570-5414, vivian_ozuna@longbeach.gov

Mary Tucker, Clean Cities Coordinator, Silicon Valley (San Jose) Clean Cities Coalition, (408) 535-8550, mary.tucker@sanjoseca.gov

Barbara Spoonhour, Clean Cities Coordinator, Northwest Riverside County Clean Cities Coalition, (951) 955-8313, spoonhour@wrcog.cog.ca.us

Burt Kronmiller, Interim Clean Cities Coordinator, Palm Springs Regional Clean Cities Coalition, 760-325-1577 x111, kronmiller@pschamber.org

Greg Newhouse, Clean Cities Coordinator, San Diego Clean Fuels Coalition, (619) 388-7673, gnewhous@sdccd.edu

Nick Haven, Acting Transportation Division Chief, Tahoe Transportation District, (775) 588-4547 x256, nhaven@trpa.org

Mike Bednarz, Clean Cities Regional Project Manager, U.S. Department of Energy, National Energy Technology Laboratory, (412) 386-4862, michael.bednarz@netl.doe.gov

Krista Fregoso, Air Pollution Specialist, California Air Resources Board, Lower Emission School Bus Program, (916) 445-5035, kfregoso@arb.ca.gov

Cherie Rainforth, California Air Resources Board, Lower Emission School Bus Program, (916) 323-2507, crainfor@arb.ca.gov

Motor Vehicle Information Hotline, California Air Resources Board, (800) 242-4450

Zero Emission Vehicle Program, California Air Resources Board, (800) 242-4450

Matt Miyasato, Air Quality Specialist, South Coast AQMD, (909) 396-3249, mmiyasato@aqmd.gov

Kathleen Mead, Manager, Retrofit Implementation Section, California Air Resources Board Mobile Source Division, (916) 324-9550, kmead@arb.ca.gov

Edie Chang, Manager - Carl Moyer Off-Road Section, California Air Resources Board, (916) 322-6924, echang@arb.ca.gov

Daniel Hawelti, Idle Reduction, California Air Resources Board, (626) 450-6149, dhawelti@arb.ca.gov

Susan Romeo, Director of Marketing and Communications, CALSTART, (626) 744-5686, sromeo@calstart.org

Jerry Wiens, Project Manager, Heavy-Duty Vehicle Programs, California Energy Commission, (916) 654-4649, jwiens@energy.state.ca.us

Peter Ward, Policy Advisor, California Energy Commission, (916) 654-4639, pward@energy.state.ca.us

Robert Chung, Deputy Director, California Transportation Commission, (916) 653-2090, robert_chung@dot.ca.gov

Fred Minassian, Technology Implementation Manager--Incentive Programs, South Coast AQMD, (909) 396-2641, fminassian@aqmd.gov

Shashi Singeetham, Air Quality Specialist, South Coast AQMD, (909) 396-3298, ssingeetham@aqmd.gov

Chuck Spagnola, Program Coordinator, San Diego APCD, (858) 586-2643, chuck.spagnola@sdcounty.ca.gov

Stan Cowen, Air Quality Engineer, Ventura County APCD, (805) 645-1408, stan@vcapcd.org

Gary Hoffman, Air Quality Engineer, Santa Barbara APCD, (805) 961-8818, gah@sbcapcd.org

Andrea Gordon, Senior Environmental Planner, Bay Area AQMD, (415) 749-4940, agordon@baaqmd.gov

Joseph Steinberger, Principal Environmental Planner, Bay Area AQMD, (415) 749-5018, jsteinberger@baaqmd.gov

Juan Ortellado, Transportation Fund for Clean Air Program, Bay Area AQMD, (415) 749-5183, jortellado@baaqmd.gov

Freya Arick, Associate Air Quality Planner/Analyst, Sacramento AQMD, Heavy-Duty Vehicle Incentive Program, (916) 874-4891, farick@airquality.org

Fleet Rule Implementation Hotline, South Coast AQMD, (909) 396-3044, fleetrules@aqmd.gov

Dean Saito, Mobile Source Strategies Manager, Technology Advancement Office, South Coast AQMD, (909) 396-3044 or (800) 288-7664, dsaito@aqmd.gov

Michelle Kirkhoff, Director of Air Quality and Mobility Programs, San Bernardino Associated Governments, (909) 884-8276 x107, mkirkhoff@sanbag.ca.gov

Department of Transportation, City of San José, (408) 535-3850, dotpermits@sanjoseca.gov

Parking Facilities Division, City of Sacramento, (916) 808-5110

Kristian Damkier, Air Quality Engineer, Sacramento Metropolitan AQMD, (916) 874-4892, kdamkier@airquality.org

Todd DeYoung, Supervising Air Quality Specialist., San Joaquin Valley Air Pollution Control District, (559) 230-5858 or (559) 230-5800, todd.deyoung@valleyair.org

Ed Huestis, Program Manager, Vacaville City Hall, (707) 449-5424, ehuestis@cityofvacaville.com

LAX Parking Services Division, (310) 646-9070

Electric Transportation Department, Sacramento Municipal Utility District, (916) 732-5283

Department of Transportation, City of Los Angeles, (213) 972-8470

Ennis Jackson, Hermosa Beach Police, (310) 318-0249, ejackson@hermoosabch.org

Lynne Taffert, Santa Monica Police Department, (310) 458-2226, lynne.taffert@smgov.net

Terry Brumgart, L.A. Department of Water and Power, (213) 367-0290

Southern California Edison, (800) 4EV-INFO
Patricia DeSpain, AFV Contact, Region 9, U.S. General Services
Administration, (928) 524-1465, patricia.despain@gsa.gov

Colorado

Incentives: None known

Regulations: All state-owned diesel vehicles must use a minimum
biodiesel blend of B20 unless it is not available or is too expensive (10
cents or more per gallon). Anyone registering a motor vehicle must report
if the vehicle is an alternative fuel vehicle or if it runs on a dedicated
alternative fuel (e.g., biodiesel exclusively, not petroleum diesel).

Contacts:

Teri Ulrich, Clean Cities Coordinator, Colorado Springs Clean
Cities Coalition, (719) 475-0155, teri@takeitforgranted.net

Denver Metro Clean Cities Coalition, cleanair@lungcolorado.org

Robin Newbrey Riesberg, Clean Cities Coordinator, Northern
Colorado Clean Cities Coalition, (970) 689-4845, cleancities@
riesberg.com

Ernie Oakes, Project Manager, U.S. Department of Energy,
Golden Field Office, (303) 275-4817, ernie.oakes@go.doe.gov

Teresa Carrillo, Commercial Vehicle Operations Manager,
Colorado Department of Transportation, (303) 757-9716, teresa.
carrillo@dot.state.co.us

Art Hale, State Fleet Manager, Colorado Department of Personnel and Administration, (303) 866-5531, art.hale@state.co.us

Tax Information Call Center, Colorado Department of Revenue,
(303) 238-7378

Stacey Simms, Governor's Biofuels Coalition, Biofuels Program
Manager, (303) 866-2401, stacey.simms@state.co.us

James Orsulak, Market Manager for Alternative Fuels, Clean Energy Fuels, (303) 322-4600, jorsulak@cleanenergyfuels.com

Walter C. Miller, Energy Services Consultant, Atmos Energy, (817) 303-2903, walter.c.miller@atmosenergy.com

Gordon Lancaster, Transportation Operations Specialist, U.S. General Services Administration, (303) 236-7599, gordon.lancaster@gsa.gov

Connecticut

Incentives: None known
Regulations: None known

Contacts:

Lee Grannis, Clean Cities Coordinator, Greater New Haven Clean Cities Coalition, Inc., (203) 627-3715, lgrannis@snet.net

Brian McGrath, Clean Cities Coordinator, Greater New Haven Clean Cities Coalition, Inc., (203) 946-7727, soggy3@aol.com

Craig Peters, Clean Cities Coordinator, Capitol Clean Cities of Connecticut, Inc., (800) 255-2631, craig.peters@manchesterhonda.com

David Levine, Clean Cities Coordinator, Capitol Clean Cities of Connecticut, Inc., (860) 653-7744, dave@ct.necoxmail.com

Ed Boman, Clean Cities Coordinator, Southwestern Area Clean Cities Coalition, (203) 256-3010, eboman@town.fairfield.ct.us

Pete Polubiatko, Clean Cities Coordinator, Norwich Clean Cities Coalition, (860) 887-6964, pete@ncdevcorp.org

Mike Scarpino, Project Manager, U.S. Department of Energy, National Energy Technology Laboratory, (412) 386-4726, michael.scarpino@netl.doe.gov

Taxpayer Services Division, Connecticut Department of Revenue, (860) 297-5962

Department of Traffic and Parking, City of New Haven, 203-946-8075

Joel M. Rinebold, Director, Energy Program, Connecticut Center for Advanced Technology, (860) 291-8832, jrinebold@ccat.us

Michael Smalec, Manager, Commercial, Industrial and Key Accounts, Southern Connecticut Gas Company/ Connecticut Natural Gas Corporation, (203) 795-7748/(860) 727-3327, msmalec@soconngas.com

Richard Guggenheim, Assistant Director, Southeastern Connecticut Council of Governments, (860) 889-2324, srguggenheim.seccog@snet.net

Robert Judge, Environmental Engineer, Region 1, U.S. Environmental Protection Agency, (617) 918-1045, judge.robert@epa.gov

Andrew E. Motter, Community Planner, U.S. Department of Transportation, Federal Transit Administration, Region 1, (617) 494-3560, andy.motter@fta.dot.gov

Delaware

Incentives: The State Energy Office's Technology Demonstration Program will provide grants for a portion of a renewable energy project's cost if that project demonstrates market potential. Such projects might include biodiesel production facilities. This program's contact person is Charlie Smisson, State Energy Coordinator, Delaware Energy Office. He can be reached at (302) 739-1530, charlie.smisson@state.de.us. Information on the state's energy programs is located at http://www.delaware-energy.com/. Furthermore, the Delaware Soybean Board provides various forms of support and assistance for biodiesel use. For additional information, visit the Board's web site at http://www.desoybeans.org or contact Jeffrey Allen, President, Delaware Soybean Board, at (302) 337-7678 or jwa913@cs.com.

Regulations: Fuel taxes are waived for U.S. or state agency vehicles that use alternative fuels. Those who own or operate vehicles that use alternative fuels must pay a special fuel tax or obtain a fuel user's license from the state's Department of Transportation. Alternative fuel suppliers are required to get a license from the state's Department of Transportation.

Contacts:

Suzanne Sebastian, Delaware Clean State Coordinator, Delaware Energy Office, (302) 735-3480, suzanne.sebastian@state.de.us

Mike Scarpino, Project Manager, U.S. Department of Energy, National Energy Technology Laboratory, (412) 386-4726, michael.scarpino@netl.doe.gov

Charlie Smisson, State Energy Coordinator, Delaware Energy Office, (302) 739-1530, charlie.smisson@state.de.us

Jeffrey Allen, President, Delaware Soybean Board, (302) 337-7678, jwa913@cs.com

Office of the Governor, (302) 744-4101 or (302) 577-3210

Ralph Schieferstein, Division Operations Manager, Chesapeake Utilities, (302) 734-6797 ext. 6734, rschieferstein@chpk.com

Andy Lambert, Vice President of Operations, SchagrinGAS, (302) 658-2000 ext. 3015, alambert@schagringas.com

Susanne Zilberfarb, Executive Director, Delaware Soybean Board, (703) 437-0995, susanne@hammondmedia.com

District of Columbia

Incentives: None known
Regulations: None known

Contacts:

George Nichols, Clean Cities Coordinator, Metropolitan Washington Council of Governments, (202) 962-3355, gnichols@mwcog.org

Kay Milewski, Project Manager, U.S. Department of Energy, National Energy Technology Laboratory, (304) 285-4535, kay.milewski@netl.doe.gov

Sabrina Williams, Energy Program Specialist, District Department of the Environment/Energy Division, (202) 671-3305, sabrina.williams@dc.gov

Rene Martinez, Manager, CNG Operations, Washington Gas, (703) 750-5830, rmartinez@washgas.com

Patricia Robinson, Administrator of Fleet Management, Department of Public Works, (202) 576-6799, patricia.robinson@dc.gov

Sylvia McMillan, Alternative Fuels Specialist, U.S. General Services Administration, (202) 619-8909, sylvia.mcmillan@gsa.gov

Florida

Incentives: Renewable energy grants are available for projects that demonstrate, commercialize, research, or develop renewable energy technologies. Materials that are used to distribute biodiesel (B10-B100), including storage, transportation, and refueling infrastructure, are exempt from a certain amount of rental, use, consumption, distribution, and storage tax. This exemption is in effect until July 1, 2010. In addition, through June 20, 2010, there is a sales and use tax credit for the costs associated with the research and development and capital operations and maintenance related to producing, storing, and distributing biodiesel (B10-B100).

Regulations: None known

Contacts:

Bill Young, Florida Space Coast Clean Cities Coalition, Clean Cities Coordinator, (321) 638-1443, (321) 638-1010, young@fsec.ucf.edu

Larry Allen, Florida Gold Coast Clean Cities Coalition, Clean Cities Coordinator, (954) 985-4416, (954) 985-4417, lallen@sfrpc.com

Steven Richardson, U.S. Department of Energy, National Energy Technology Laboratory, Project Manager, (304) 285-4185, (304) 285-4638, steven.richardson@netl.doe.gov

Florida Energy Office, General Inquiries, (850) 245-8002

Florida Division of Motor Vehicles, (850) 922-9000, dmv@hsmv.state.fl.us

Jill Stoyshich, Florida Energy Office, Manager, Hydrogen Program, (850) 245-8277, (850) 245-8003, jill.stoyshich@dep.state.fl.us

James Culp, Technology Research and Development Authority, Energy Programs Manager, (321) 269-6330, (321) 383-5260, jculp@trda.org

Dale Aspy, U.S. Environmental Protection Agency, Environmental Engineer, Region 4 Air Planning Branch, (404) 562-9041, (404) 562-9019, aspy.dale@epa.gov

Wes Allen, U.S. General Services Administration, Transportation Specialist, Southeast Region, (404) 331-3052, (404) 331-1879, james.allen@gsa.gov

Georgia

Incentives: None known

Regulations: When available and economically feasible, all state agencies and departments must maximize the use of biodiesel blends and state agency vehicles. All biodiesel that is made or sold in Georgia must meet the ASTM D 6751 standard.

Contacts:

Wendy Morgan, Clean Cities Co-Coordinator, Atlanta Clean Cities Coalition, (678) 244-4152, wendy@cte.tv

Charise Stephens, Clean Cities Director, Middle Georgia Clean Cities Coalition, (478) 751-9178 or (478) 751-9101, charise.stephens@macon.ga.us

Steven Richardson, Project Manager, U.S. Department of Energy, National Energy Technology Laboratory, (304) 285-4185, steven.richardson@netl.doe.gov

James Udi, Environmental Specialist, Georgia Environmental Protection Division, (404) 363-7046, james_udi@dnr.state.ga.us

Gerald Ross, Division Director of Transportation Data and Intermodal Development, Georgia Department of Transportation, (404) 656-0610, gerald.ross@dot.state.ga.us

Ben Echols, Product Manager- Electric Mobility, Georgia Power Company, (404) 506-6713, bdechols@southernco.com

Dale Aspy, Environmental Engineer, Region 4 Air Planning Branch, U.S. Environmental Protection Agency, (404) 562-9041, aspy.dale@epa.gov

Alan Powell, Environmental Engineer, Region 4 Air Planning Branch, U.S. Environmental Protection Agency, (404) 562-9045, powell.alan@epa.gov

Walter C. Miller, Energy Services Consultant, Atmos Energy, (817) 303-2903, walter.c.miller@atmosenergy.com

Wes Allen, Transportation Specialist, Southeast Region, U.S. General Services Administration, (404) 331-3052, james.allen@gsa.gov

Hawaii

Incentives: The Department of Agriculture's Energy Feedstock Program encourages the production of feedstock for biofuels. State agencies must give preference to biofuels or biofuel blends.

Regulations: Of course, biodiesel must meet the ASTM specification. The state must provide support for the development of alterative fuels. Biodiesel distributors must pay a license tax for every gallon sold or used by the distributor if the biodiesel will be used on state public highways in an internal combustion engine.

Contacts:

Robert Primiano, Clean Cities Coordinator, Honolulu Clean Cities Coalition, (808) 768-3500, rprimiano@honolulu.gov

Mike Bednarz, Project Manager, U.S. Department of Energy, National Energy Technology Laboratory, (412) 386-4862, michael.bednarz@netl.doe.gov

Maria Tome, Alternate Energy Engineer, Hawaii Department of Business, Economic Development, and Tourism, Strategic Industries Division, (808) 587-3809, mtome@dbedt.hawaii.gov

Hawaii Department of Business, Economic Development, and Tourism, (808) 587-3814

Hoku Keolanui, Account Executive, The Gas Company, LLC, (808) 594-5585, tkeolanui@hawaiigas.com

Hawaii State Department of Taxation, (800) 222-3229

Collette Craig, AFV Contact, Region 9, U.S. General Services Administration, (928) 524-3975, collette.craig@gsa.gov

Idaho

Incentives: Until December 31, 2011, there is a tax credit for qualified biofuel refueling infrastructure. There also is a biofuel infrastructure grant fund, which provides funds for a portion of the cost associated with the installation of new biofuel infrastructure. Finally, licensed distributors can get a tax deduction based on the renewable content of the fuel that they sell.

Regulations: There is an excise tax on biodiesel, but the motor fuel tax rate does not apply.

Contacts:

Sandy Shuptrine, Clean Cities Coordinator/Executive Director, Greater Yellowstone/Teton Clean Energy Coalition, (307) 733-6371, sandyshuptrine@wyom.net

Beth Baird, Clean Cities Coordinator, Treasure Valley Clean Cities Coalition, (208) 384-3984, bbaird@cityofboise.org

Ernie Oakes, Project Manager, U.S. Department of Energy, Golden Field Office, (303) 275-4817, ernie.oakes@go.doe.gov

John Crockett, Energy Specialist, Idaho Energy Division, (208) 287-4894, john.crockett@idwr.idaho.gov

Jackie McCloughan, Fuel Systems Manager, Idaho Transportation Department, (208) 334-8094, jackie.mccloughan@itd.idaho.gov

Gordon Larsen, Natural Gas Vehicle Supervisor, Questar Gas, (801) 324-5987, gordon.larsen@questar.com

Jim Grambihler, Natural Gas Vehicle Operations, Questar Gas, (801) 324-5119, jim.grambihler@questar.com

Julie Shain, Fleet Manager, U.S. General Services Administration, (208) 321-9150, julie.shain@gsa.gov

Illinois

Incentives: The state's Department of Commerce and Economic Opportunity runs the Renewable Fuels Research, Development, and Demonstration Program. This program works to promote and increase biofuels use and provides grants for the development of business plans, studies, and other work associated with new Illinois-based biofuels facilities. For more information, contact Dave Loos, Illinois Department of Commerce and Economic Opportunity, Illinois State Energy Office, at (217) 785-3969, dave.loos@illinois. gov. For information on the program, visit http://www.commerce. state.il.us/dceo/Bureaus/Energy_Recycling/.

Alternative fuel rebates also are available through the Illinois Alternate Fuels Rebate Program. For additional information, contact Darwin Burkhart, Manager, Mobile Source Programs, Illinois Environmental Protection Agency, at (217) 557-1441, darwin.burkhart@ epa.state.il.us, or visit the web site at http://www.epa.state.il.us/air/.

Regulations: A tax exemption is provided for a portion of biodiesel and biodiesel blend sales and use taxes. Moreover, state and local government entities must use a minimum of B2 when refueling at a bulk central fueling facility. State agencies are allowed to give preference in awarding contracts to proposals that will use biodiesel made from Illinois soybeans. The state is mandated to do all it can to facilitate the purchase of B2 blends for its diesel vehicle fleet. In order to support the state's use of biofuels, the governor's office has established a goal of supplying 50% of its own energy from homegrown biofuels by 2017.

Contacts:

Bethany Kraseman, Clean Cities Coordinator, Chicago Area Clean Cities Coalition, (773) 320-1718, bethany.kraseman@chicagocleancities.org

Mike Scarpino, Project Manager, U.S. Department of Energy, National Energy Technology Laboratory, (412) 386-4726, michael.scarpino@netl.doe.gov

Dave Loos, Illinois Department of Commerce and Economic Opportunity, Illinois State Energy Office, (217) 785-3969, dave.loos@illinois.gov

Darwin Burkhart, Manager, Mobile Source Programs, Illinois Environmental Protection Agency, (217) 557-1441, darwin.burkhart@epa.state.il.us

Walter C. Miller, Energy Services Consultant, Atmos Energy, (817) 303-2903, walter.c.miller@atmosenergy.com

Scott Benson, Transportation Specialist, Great Lakes Region, U.S. General Services Administration, (312) 886-8682, scott.benson@gsa.gov

Indiana

Incentives: The Economic Development Corporation has developed the Indiana Twenty-First Century Research and Technology Fund through which grants and loans will be provided for alternative fuel technologies. Furthermore, tax credits are available for biodiesel producers, blenders, and retailers. Additionally, government entities that use biodiesel for essential government services are entitled to a 10% discount on biodiesel blends of B20 by volume or higher.

Regulations: State bodies and educational institutions are required to purchase biofuels whenever possible; all diesel vehicles must use a B2 blend or higher. In addition, biodiesel blends of B20 and higher that are used by individuals for personal, non-commercial reasons are exempt from the license tax.

Contacts:

Kellie Walsh, Executive Director, Central Indiana Clean Cities Alliance, Inc., (317) 834-3754, klwcicca@aol.com

Carl Lisek, Clean Cities Coordinator, South Shore Clean Cities, Inc., (219) 365-4289, southscc@comcast.net

Mike Scarpino, Project Manager, U.S. Department of Energy, National Energy Technology Laboratory, (412) 386-4726, michael. scarpino@netl.doe.gov

Barbara Pesut-Hanley, Industrial Sales Consultant, Citizens Gas & Coke Utility, (317) 927-4431, bpesut-hanley@cgcu.com

Cary Aubrey, Bioenergy Program Manager, Indiana State Department of Agriculture, (317) 450-0652, caubrey@isda.in.gov

Scott Deloney, Section Chief, Program Planning and Policy, Indiana Clean Fuels Fleet Program, (317) 233-5684, sdeloney@ idem.in.gov

Fuel Tax Section, Indiana Department of Revenue, (317) 615-2630

Scott Benson, Transportation Specialist, Great Lakes Region, U.S. General Services Administration, (312) 886-8682, scott. benson@gsa.gov

Iowa

Incentives: The Iowa Power Fund offers support for the research, development, commercialization, and deployment of biofuels. Additionally, financial incentives are offered for the conversion or installation of infrastructure needed to establish biodiesel distribution terminals at service stations; this program is scheduled to end on June 30, 2008. A tax credit is provided to biodiesel retailers for whom biodiesel makes up at least half of their petroleum diesel sales.

Grants also are available to biodiesel blenders through the state's Renewable Fuel Infrastructure Program. Those interested in this biofuels infrastructure grant program can contact Dick Vegors, Marketing Manager, Grain and Grain Co-Products, Iowa Department of Economic Development, Business Development Division, Domestic & International, at (515) 242-4796, dick.vegors@iowalifechanging.com. Additional program information is available at www.iowalifechanging.com/business/renewablefuels.html.

Alternative fuel loans are available through the Alternative Energy Revolving Loan Program. The contact person for this program is Keith Kutz, Administrative Specialist, Iowa Energy Center. He can be reached at (515) 294-8819, iec@energy.iastate.edu. Information on the loan program is available at http://www.energy.iastate.edu/funding/aerlp-index.html.

Another alternative fuel loan is available through the state's Value-Added Agricultural Products and Processes Financial Assistance Program. Projects that focus on research and development are not funded by this program. For additional information, contact the Business Finance, Business Development Division, Program Coordinator of the Iowa Department of Economic Development, Business Development Division. The phone number is (515) 242-4819, and the email address is business@iowalifechanging.com. Additional information can be viewed at http://www.iowalifechanging.com.

Tax credits for the commercial production of alternative fuels are available through the High Quality Job Creation Program and the Enterprise Zone Program. For more information, contact the Business Finance, Business Development Division, Program Coordinator of the Iowa Department of Economic Development, Business Development Division. The phone number is (515) 242-4819, and the email address is business@iowalifechanging.com. Additional information can be viewed at http://www.iowalifechanging.com.

Regulations: Retailers are required to blend a certain percentage of biodiesel with petroleum diesel. In all of their bulk petroleum diesel fuel purchases, all state agencies must include a minimum 5% biodiesel by 2007, 10% biodiesel by 2008, and 20% biodiesel by 2010. All biodiesel must meet the ASTM specification. The state treasury's biodiesel fuel revolving fund will be used to purchase biodiesel for the state's Department of Transportation diesel vehicles; such vehicles will display a sticker indicating that they use biodiesel.

Contacts:

Brian Crowe, Clean Cities Coordinator, Iowa Clean Cities Coalition, (515) 281-8518, brian.crowe@dnr.state.ia.us

Neil Kirschner, Project Manager, U.S. Department of Energy, National Energy Technology Laboratory, (412) 386-5793, neil.kirschner@netl.doe.gov

Keith Kutz, Administrative Specialist, Iowa Energy Center, (515) 294-8819, iec@energy.iastate.edu

Business Finance, Business Development Division, Program Coordinator, Iowa Department of Economic Development, Business Development Division, (515) 242-4819, business@iowalifechanging.com

Dick Vegors, Marketing Manager, Grain and Grain Co-Products, Iowa Department of Economic Development, Business Development Division, Domestic & International, (515) 242-4796, dick.vegors@iowalifechanging.com

Lucy Norton, Iowa Renewable Fuels Association, (515) 252-6249, info@iowarfa.org

Gene Jones, Transportation Planner, Iowa Department of Transportation, Office of Program Management, (515) 239-1054, g.jones@dot.state.ia.us

Alan Banwart, Environmental Protection Specialist, Region 7, U.S. Environmental Protection Agency, (913) 551-7819, banwart.alan@epa.gov

Joan Roeseler, Federal Transit Administration, Region 7, U.S. Department of Transportation, (816) 329-3936, joan.roeseler@dot.gov

Don Gard, Transportation Operations Specialist, U.S. General Services Administration, Regional Fleet Management Office, (816) 823-3625, don.gard@gsa.gov

Kansas

Incentives: Biodiesel producers can get an incentive for production of 30 cents per gallon of biodiesel sold. Additionally, licensed retailers of motor fuels can receive an incentive for selling biodiesel and other renewable fuels. For information, contact Cindy Mongold, Public Service Administrator II, Kansas Department of Revenue, (785) 296-7048, cindy_mongold@kdor.state.ks.us, or visit the agency's web site at http://www.ksrevenue.org.

An alternative fuel vehicle income tax credit is available to vehicles that run on biodiesel and other fuels. There also is an alternative fuel refueling infrastructure tax credit for refueling stations. For information about either of these tax credits, contact Jim Ploger, Director of Renewable Energy & Energy Efficiency, Kansas Energy Office, at (785) 271-3349, j.ploger@kcc.state.ks.us, or visit the agency's web site at http://www.kcc.state.ks.us/energy/alt_fuels.htm.

Regulations: All state-owned diesel vehicles and equipment must use a biodiesel blend of B2 or higher. Exceptions are made when biodiesel is not available or when it is too costly.

Contacts:

Bob Housh, Kansas City Regional Clean Cities Coalition, Interim Clean Cities Coordinator, (816) 531-7283 or (877) 620-1803, housh@kcenergy.org

Neil Kirschner, U.S. Department of Energy, National Energy Technology Laboratory, Project Manager, (412) 386-5793, neil.kirschner@netl.doe.gov

Jim Ploger, Kansas Energy Office, Director of Renewable Energy & Energy Efficiency, (785) 271-3349, j.ploger@kcc.state.ks.us

Patricia Platt, Kansas Department of Revenue, Public Service Administrator II, (785) 291-3670, patricia_platt@kdor.state.ks.us

Cindy Mongold, Kansas Department of Revenue, Public Service Administrator II, (785) 296-7048, cindy_mongold@kdor.state.ks.us

Tom Whitaker, Kansas Motor Carriers Association, Executive Director, (785) 267-1641, tomw@kmca.org

Joan Roeseler, U.S. Department of Transportation, Federal Transit Administration, Region 7, (816) 329-3936, joan.roeseler@fta.dot.gov

Don Gard, U.S. General Services Administration, Regional Fleet Management Office, Transportation Operations Specialist, (816) 823-3625, don.gard@gsa.gov

Alan Banwart, U.S. Environmental Protection Agency, Environmental Protection Specialist, Region 7, (913) 551-7819, banwart.alan@epa.gov

Kentucky

Incentives: Biodiesel producers and blenders can get an income tax credit of $1.00/gallon for blends of B2 and higher.

Regulations: The Commonwealth's Office of Energy Policy is required to create and implement a strategy toward energy independence, which includes the use of biodiesel and other biofuels. Employees of the Kentucky Transportation Cabinet must use a blend of B2 in their diesel vehicles whenever possible.

Contacts:

Melissa Howell, Commonwealth Clean Cities Partnership, Inc., Clean Cities Coordinator, (502) 452-9152 or (502) 593-3846, (502) 452-9152, kcfc@aol.com

Kay Milewski, U.S. Department of Energy, National Energy Technology Laboratory, Project Manager, (304) 285-4535, kay. milewski@netl.doe.gov

James Bush, Governor's Office of Energy Policy, Division of Renewable Energy and Energy Efficiency, (502) 564-7192, (502) 564-7484, james.bush@ky.gov

John T. (Tom) Underwood, Kentucky Propane Education and Research Council, Executive Director, (502) 223-5322, (502) 223-4937, info@choosepropane.org

Lynn Sopowroski, Kentucky Transportation Cabinet, Transportation Engineer Branch Manager, (502) 564-7183, (502) 564-2865, lynn.soporowski@ky.gov

Jesse Mayes, Kentucky Transportation Cabinet, Transportation Engineering Specialist for Air Quality, (502) 564-7183, (502) 564-2865, jesse.mayes@ky.gov

Walter C. Miller, Atmos Energy, Energy Services Consultant, (817) 303-2903, (817) 303-2929, walter.c.miller@atmosenergy.com

Wes Allen, U.S. General Services Administration, Transportation Specialist, Southeast Region, (404) 331-3052, (404) 331-1879, james.allen@gsa.gov

Alan Powell, U.S. Environmental Protection Agency, Environmental Engineer, Region 4 Air Planning Branch, (404) 562-9045, (404) 562-9019, powell.alan@epa.gov

Dale Aspy, U.S. Environmental Protection Agency, Environmental Engineer, Region 4 Air Planning Branch, (404) 562-9041, (404) 562-9019, aspy.dale@epa.gov

Louisiana

Incentives: Louisiana offers a biodiesel equipment and fuel tax exemption for some types of property and equipment that are used in the production or extraction of biodiesel.

Regulations: Once the state's annual biodiesel production reaches or surpasses 10 million gallons, all petroleum diesel fuel sold in Louisiana must be a biodiesel blend of B2 within six months. This biodiesel must be made using US-grown feedstock. All renewable fuel plants in the state that make biodiesel using crops must use a minimum of 2.5% of Louisiana's soybean crop as feedstock.

Contacts:

Tammy Morgan, Greater Baton Rouge Clean Cities Coalition, Clean Cities Coordinator, (225) 389-8560, tlmorgan@brgov.com

Vicki Cappel, New Orleans Reg. Planning Commission/Greater New Orleans Clean Cities Coalition (Not Yet Designated, Clean Cities Coordinator, (504) 568-6627, (504) 568-6643, vcappel@ norpc.org

Wes Wyche, City of Shreveport/Greater Shreveport Clean Cities Coalition (Not Yet Designated), Department of Operational Services/Clean Cities Coordinator, (318) 673-6072, (318) 673-7663, wes.wyche@ci.shreveport.la.us

Neil Kirschner, U.S. Department of Energy, National Energy Technology Laboratory, Project Manager, (412) 386-5793, neil. kirschner@netl.doe.gov

J. Bryan Crouch, Louisiana Department of Natural Resources, Engineer, Alternative Fuels and Refinery , (225) 342-2122, (225) 242-3605, john.crouch@la.gov

Louisiana Department of Revenue, Taxpayer Services Division, (225) 219-0102

Louisiana Department of Natural Resources, (225) 342-1399

Robert Borne, Entergy Corporation, Business Development, (225) 763-5117, (225) 763-5168, rborne@entergy.com

Walter C. Miller, Atmos Energy, Energy Services Consultant, (817) 303-2903, (817) 303-2929, walter.c.miller@atmosenergy.com

Gordon Lancaster, U.S. General Services Administration, Transportation Operations Specialist, (303) 236-7599, (303) 236-7590, gordon.lancaster@gsa.gov

Sandra Rennie, U.S. Environmental Protection Agency, Mobile Source Team Leader, Region 6, (214) 665-7367, (214) 665-7263, rennie.sandra@epa.gov

Maine

Incentives: Commercial biodiesel producers can get a state income tax credit for biodiesel to be used in motor vehicles or that would be a substitute for liquid fuels. Additionally, those who construct, install, or improve upon biodiesel refueling stations might obtain a tax credit through December 31, 2008.

Regulations: Loan and subsidies are available through the Agriculturally Derived Fuel Fund to businesses or cooperatives that design and build biofuels.

Contacts:

Steve Linnell, Clean Cities Coordinator, Maine Clean Communities , (207) 774-9891, slinnell@gpcog.org

Mike Scarpino, Project Manager, U.S. Department of Energy, National Energy Technology Laboratory, (412) 386-4726, michael. scarpino@netl.doe.gov

Lynne Cayting, Bureau of Air Quality, Mobile Sources Section Chief, Maine Department of Environmental Protection, (207) 287-7599, lynne.a.cayting@maine.gov

John Duncan, Director, Portland Area Comprehensive Transportation Committee (MPO), (207) 774-9891, jduncan@gpcog.org

Denis Bergeron, Director, Energy Conservation Division, Maine Public Utilities Commission, (207) 287-1366, denis.bergeron@ maine.gov

Andrew E. Motter, Community Planner, U.S. Department of Transportation, Federal Transit Administration, Region 1, (617) 494-3560, andy.motter@dot.gov

Robert Judge, Environmental Engineer, Region 1, U.S. Environmental Protection Agency, (617) 918-1045, judge.robert@epa.gov

Maryland

Incentives: Biodiesel producers may receive credits for biodiesel production by applying to the Renewable Fuels Incentive Board.

Regulations: Starting in fiscal year 2008, at least half of the state's diesel vehicles must use a biodiesel blend of B5 or higher.

Contacts:

Chris Rice, Maryland Clean Cities Coordinator, Maryland Energy Administration, (410) 260-7207, crice@energy.state.md.us

George Nichols, Washington Metropolitan Clean Cities Coordinator, Metropolitan Washington Council of Governments, (202) 962-3355, gnichols@mwcog.org

Steven Richardson, Project Manager, U.S. Department of Energy, National Energy Technology Laboratory, (304) 285-4535, steven. richardson@netl.doe.gov

Susanne Zilberfarb, Biodiesel Project Leader, Maryland Soybean Board, 410) 742-9500 or (703) 437-0995, shammond@ezy.net

Howard Simmons, Manager-Air Quality Programs, Maryland Department of Transportation, (410) 865-1296, hsimmons@mdot. state.md.us

Tim Shepherd, Division Chief, Mobile Sources Control Program, Maryland Department of the Environment, Air and Radiation Management Administration, (410) 537-3236, tshepherd@mde. state.md.us

Sylvia McMillan, Alternative Fuel Coordinator, U.S. General Services Administration, (202) 619-8992, sylvia.mcmillan@gsa.gov

Massachusetts

Incentives: None known

Regulations: None known

Contacts:

David Rand, Clean Cities Coordinator, Massachusetts Clean Cities Coalition, (617) 727-4732 x40138, david.rand@state.ma.us

Mike Scarpino, Project Manager, U.S. Department of Energy, National Energy Technology Laboratory, (412) 386-4726, michael. scarpino@netl.doe.gov

Mike Manning, Lead Account Executive - NGVs, Keyspan Energy Delivery, (781) 466-5373, mmanning@keyspanenergy.com

Andrew E. Motter, Community Planner, U.S. Department of Transportation, Federal Transit Administration, Region 1, (617) 494-3560, andy.motter@fta.dot.gov

Robert Judge, Environmental Engineer, Region 1, U.S. Environmental Protection Agency, (617) 918-1045, judge.robert@epa.gov

Michigan

Incentives: Those who use a biodiesel blend of B5 or higher can claim a reduced biofuels tax. Of course, this fuel must meet the ASTM specification for biodiesel. Moreover, there is a property tax exemption for those who use industrial property to make biodiesel. Grants also are available to refueling stations that offer biodiesel blends.

Regulations: Biodiesel blenders or suppliers that are not part of the bulk transfer terminal system are required to get a blender's license and must meet reporting requirements. Of course, biodiesel sold in Michigan must meet the state's quality requirements.

Contacts:

Dan Radomski, Clean Cities Coordinator, Detroit Clean Cities Coalition/NextEnergy, (313) 833-0100 x150, detroitcleancities@nextenergy.org

David Konkle, Clean Cities Coordinator, Ann Arbor Clean Cities Coalition, (734) 996-3150, dkonkle@ci.ann-arbor.mi.us

Murray Britton, GLACCC Board Chairman, Greater Lansing Area Clean Cities Coalition, (517) 483-4465, glaccc@hotmail.com

Mike Scarpino, Project Manager, U.S. Department of Energy, National Energy Technology Laboratory, (412) 386-4726, michael.scarpino@netl.doe.gov

Tim Shireman, Department of Labor and Economic Growth, Michigan Energy Office, (517) 241-6281, tashire@michigan.gov

Jody Pollok, Executive Director, Corn Marketing Program of

Michigan & Michigan Corn Growers Association, (517) 668-2676, jpollok@micorn.org

Crystal Bollman, Communication and Programs Coordinator, Corn Marketing Program of Michigan and Michigan Corn Growers Association, (517) 668-2676, cbollman@micorn.org

Pete Porciello, Air Quality Specialist, Michigan Department of Transportation, (517) 335-2603, porciellop@michigan.gov

Robert Rusch, Environmental Quality Specialist, Strategic Development Unit, Michigan Department of Environmental Quality, (517) 373-7041, ruschr@michigan.gov

Teresa Walker, Senior Environmental Quality Analyst, Emissions Trading Programs, Michigan Department of Environmental Quality, (517) 335-2247, walkertr@michigan.gov

Scott Benson, Transportation Specialist, Great Lakes Region, U.S. General Services Administration, (312) 886-8682, scott.benson@gsa.gov

Minnesota

Incentives: None known

Regulations: All state agencies must work to enhance the infrastructure for biodiesel and to use biodiesel blends (B20-B100) in state diesel vehicles whenever it is feasible. All petroleum diesel fuel sold in Minnesota must contain a minimum biodiesel blend of B2.

Contacts:

Tim Gerlach, Clean Cities Coordinator, American Lung Association of the Upper Midwest, (651) 223-9577, gerlach@alamn.org

Mike Scarpino, Project Manager, U.S. Department of Energy, National Energy Technology Laboratory, (412) 386-4726, michael.scarpino@netl.doe.gov

John Scharffbillig, Fleet Manager, Minnesota Department of Transportation, (612) 725-2354, john.scharffbillig@dot.state.mn.us

Ralph Groschen, Senior Marketing Specialist, Minnesota Department of Agriculture, (651) 201-6223, ralph.groschen@state.mn.us

Michael Bull, Assistant Commissioner for Renewable Energy, Minnesota Department of Commerce, (651) 282-5011, michael.bull@state.mn.us

Scott Benson, Transportation Specialist, Great Lakes Region, U.S. General Services Administration, (312) 886-8682, scott.benson@gsa.gov

Mississippi

Incentives: Biodiesel producers can be paid by the Commissioner of Agriculture and Commerce for producing biodiesel.

Regulations: The Department of Finance and Administration will promote the use of biodiesel and other alternative fuels to state agencies.

Contacts:

Steven Richardson, Project Manager, U.S. Department of Energy, National Energy Technology Laboratory, (304) 285-4185, steven.richardson@netl.doe.gov

Mitch Ferrell, Director, Mississippi Department of Insurance, Liquefied Compressed Gas Division, (601) 359-1064, mitch.ferrell@mid.state.ms.us

Larissa Williams, Utilization Engineer, Atmos Energy, (601) 961-6964, larissa.williams@atmosenergy.com

Motice Bruce, Bureau/Set Manager, Mississippi Development Authority Energy Division, (601) 359-6600, mbruce@mississippi.org

Dale Aspy, Environmental Engineer, Region 4 Air Planning Branch, U.S. Environmental Protection Agency, (404) 562-9041, aspy.dale@epa.gov

Alan Powell, Environmental Engineer, Region 4 Air Planning Branch, U.S. Environmental Protection Agency, (404) 562-9045, powell.alan@epa.gov

Wes Allen, Transportation Specialist, Southeast Region, U.S. General Services Administration, (404) 331-3052, james.allen@gsa.gov

Missouri

Incentives: Monthly grants are available to biodiesel producers through the Missouri Qualified Biodiesel Producer Incentive Fund. For more information, contact Robin Perso, Director of Budget and Planning, Missouri Department of Agriculture, at (573) 526-4892, robin.perso@mda.mo.gov.

School districts can work with farmer-owned, non-profit cooperatives to purchase biodiesel blends of B20 or higher.

Regulations: The Missouri Department of Transportation can use biodiesel blends of B20 or higher in its diesel vehicle fleet and equipment.

Contacts:

Kevin Herdler, Clean Cities Coordinator, St. Louis Clean Cities Program, (314) 822-5831, herdlekc@kirkwoodmo.org

Bob Housh, Clean Cities Coordinator, Kansas City Regional Clean Cities Coalition, (816) 531-7283, housh@kcenergy.org

Neil Kirschner, Project Manager, U.S. Department of Energy, National Energy Technology Laboratory, (412) 386-5793, neil.kirschner@netl.doe.gov

Cindy Carroll, Energy Specialist, Missouri Department of Natural Resources, Missouri Energy Center, (573) 751-3443, cindy. carroll@dnr.mo.gov

Robin Perso, Director of Budget and Planning, Missouri Department of Agriculture, (573) 526-4892, robin.perso@mda.mo.gov

Steve Nagle, Director of Planning, East-West Gateway Council of Governments, (314) 421-4220, steve.nagle@ewgateway.org

Jeannie Wilson, Fleet Manager, Missouri Department of Transportation, (573) 526-1199, jeannie.wilson@modot.mo.gov

Becky Grisham, Director of Communications, Missouri Corn Growers Association, (573) 893-4181, bgrisham@mocorn.org

Tom Schultz, Natural Gas Vehicle Market Development Manager, Laclede Gas Company, (314) 342-0684, tschultz@lacledegas.com

Bob Noelker, Project Engineering Manager, Laclede Gas Company, (314) 658-5594, bnoelker@lacledegas.com

Walter C. Miller, Energy Services Consultant, Atmos Energy, (817) 303-2903, walter.c.miller@atmosenergy.com

Joan Roeseler, Federal Transit Administration, Region 7, U.S. Department of Transportation, (816) 329-3936, joan.roeseler@ fta.dot.gov

Don Gard, Transportation Operations Specialist, U.S. General Services Administration, Regional Fleet Management Office, (816) 823-3625, don.gard@gsa.gov

Alan Banwart, Environmental Protection Specialist, Region 7, U.S. Environmental Protection Agency, (913) 551-7819, banwart. alan@epa.gov

Montana

Incentives: Those involved in researching, producing, or developing biodiesel might be eligible for a property tax incentive. In addition, individuals and businesses can get a tax credit for part of the cost of storing and blending biodiesel. A tax credit also is available for a portion of the cost of biodiesel production facilities, and there is a biodiesel production tax incentive that is available to producers during their first three years of production.

Regulations: None known

Contacts:

Sandy Shuptrine, Clean Cities Coordinator/Executive Director, Greater Yellowstone/Teton Clean Energy Coalition, (307) 733-6371, sandyshuptrine@wyom.net

Ernie Oakes, Project Manager, U.S. Department of Energy, Golden Field Office, (303) 275-4817, ernie.oakes@go.doe.gov

Howard Haines, State Bioenergy Program Manager, Montana Department of Environmental Quality, (406) 841-5252, hhaines@mt.gov

Montana Hydrogen Futures Project, The University of Montana - Missoula College of Technology, Dr. R. Paul Williamson, University of Montant-Missoula College of Technology, 909 South Ave. West Missoula, MT 59801, (406) 243-7851, paul.williamson@umontana.edu

Gordon Lancaster, Transportation Operations Specialist, U.S. General Services Administration, (303) 236-7599, gordon.lancaster@gsa.gov

Nebraska

Incentives: None known

Regulations: All state employees who use diesel vehicles must use biodiesel blends whenever it is feasible to do so. The state is working to increase the availability of B2 for state employees who use diesel vehicles. Additionally, some motor fuels are exempt from motor fuel tax laws when sold to or manufactured at a biodiesel production facility.

Contacts:

Neil Kirschner, Project Manager, U.S. Department of Energy, National Energy Technology Laboratory, (412) 386-5793, neil. kirschner@netl.doe.gov

General Inquiries, Nebraska State Energy Office, (402) 471-2867, energy1@mail.state.ne.us

Jerry Loos, Public Information Officer, Nebraska State Energy Office, (402) 471-3356, jloos@neo.ne.gov

Tom Sands, Operations Division Manager, Nebraska Department of Roads, (402) 479-4339, tsands@dor.state.ne.us

Michael Mattison, Maintenance Engineer, Nebraska Department of Roads, (402) 479-4878, michaelmattison@dor.state.ne.us

Joan Roeseler, Federal Transit Administration, Region 7, U.S. Department of Transportation, (816) 329-3936, joan.roeseler@fta.dot.gov

Alan Banwart, Environmental Protection Specialist, Region 7, U.S. Environmental Protection Agency, (913) 551-7819, banwart. alan@epa.gov

Don Gard, Transportation Operations Specialist, U.S. General Services Administration, Regional Fleet Management Office, (816) 823-3625, don.gard@gsa.gov

Nevada

Incentives: None known
Regulations: None known

Contacts:

Dan Hyde, Clean Cities Coordinator, Las Vegas Inc. Clean Cities Coalition, (702) 229-6971, dhyde@lasvegasnevada.gov

Duane Sikorski, Clean Cities Coordinator, Truckee Meadows Inc. Clean Cities Coalition, (775) 784-7206, dsikorsk@washoecounty.us

Nick Haven, Principal Transportation Planner, Tahoe Transportation District, (775) 588-4547 x256, nhaven@trpa.org

Mike Bednarz, Project Manager, U.S. Department of Energy, National Energy Technology Laboratory, (412) 386-4862, michael.bednarz@netl.doe.gov

Peter Konesky, Energy Specialist, Nevada State Energy Office, (775) 684-8735, pkonesky@dbi.state.nv.us

Russ Law, Chief Operations Analysis Engineer, Nevada Department of Transportation, (775) 888-7192, rlaw@dot.state.nv.us

Patricia DeSpain, AFV Contact, Region 9, U.S. General Services Administration, (928) 524-1465, patricia.despain@gsa.gov

New Hampshire

Incentives: None known
Regulations: The state's Department of Transportation has undertaken a biodiesel pilot program in which biodiesel is used by some diesel vehicles operated by the agency and the University of New Hampshire.

Contacts:

Becky Ohler, Granite State Clean Cities Coordinator, New Hampshire Department of Environmental Services, (603) 271-6749, rohler@des.state.nh.us

Barbara Bernstein, Granite State Clean Cities Co-Coordinator, New Hampshire Department of Environmental Services, (603) 274-6751, bbernstein@des.state.nh.us

Mike Scarpino, Project Manager, U.S. Department of Energy, National Energy Technology Laboratory, (412) 386-4726, michael.scarpino@netl.doe.gov

New Hampshire Department of Environmental Services, (603) 271-3503

Robert Judge, Environmental Engineer, Region 1, U.S. Environmental Protection Agency, (617) 918-1045, judge.robert@epa.gov

New Jersey

Incentives: A biodiesel use rebate is available for government authorities, local governments, school districts, and state colleges and universities that are eligible. For additional information, contact Ellen Bourbon, Clean Cities Coordinator, New Jersey Clean Cities Coalition, (609) 984-3058, ellen.bourbon@bpu.state.nj.us.

Regulations: All buses that the New Jersey Transit Corporation purchases must run on biodiesel or another cleaner fuel.

Contacts:

Mike Scarpino, Project Manager, U.S. Department of Energy, National Energy Technology Laboratory, (412) 386-4726, michael.scarpino@netl.doe.gov

Dick Duffy, Gas Products Manager, Public Service Electric and Gas Company, (973) 430-7664, richard.duffy@pseg.com

Reema Loutan, Environmental Engineer, Region 2, U.S. Environmental Protection Agency, (212) 637-3760, loutan.reema@epa.gov

Tristan Gillespie, Pollution Prevention Coordinator, Region 2, U.S. Environmental Protection Agency, (212) 637-3753, gillespie.tristan@epa.gov

New Mexico

Incentives: New Mexico offers a biodiesel income tax credit to blending facilities for blends B2 and higher.

Regulations: All public schools, state agencies, and state political subdivisions must use a biodiesel blend of B5 or higher in diesel vehicles. This law goes into effect on July 1, 2010. After July 1, 2012, all of the petroleum diesel fuel sold in New Mexico must contain a minimum blend of B5. This requirement might be suspended if price or availability makes it difficult to achieve. By 2010, all public schools, higher education institutions, and cabinet-level state agencies must get at least 15% of their fuel from renewable sources, such as biodiesel.

Contacts:

Frank Burcham, Clean Cities Coordinator/ Executive Director, Land of Enchantment Clean Cities Coalition/ Alternative Fuels Vehicle Network (AFVN), (505) 856-8585, loecleancities@comcast.net

Neil Kirschner, Project Manager, U.S. Department of Energy, National Energy Technology Laboratory, (412) 386-5793, neil.kirschner@netl.doe.gov

Richard Leonard, Alternative Fuels Program Manager, Energy, Mineral, and Natural Resources Department, (505) 476-3316, rleonard@state.nm.us

Mark Sprick, Transportation Planning Services Manager, Mid-Region Council of Governments, (505) 247-1750, msprick@mrcog-nm.gov

Louise Martinez, Bureau Chief, Energy, Minerals, and Natural Resources Department, (505) 476-3315, louise.n.martinez@state.nm.us

Colin Messer, Program Manager, Energy, Mineral, and Natural Resources Department, Energy Conservation and Management Division, (505) 476-3314, colinj.messer@state.nm.us

James Orsulak, Market Manager for Alternative Fuels, Clean Energy Fuels, (303) 322-4600, jorsulak@cleanenergyfuels.com

Sandra Rennie, Mobile Source Team Leader, Region 6, U.S. Environmental Protection Agency, (214) 665-7367, rennie.sandra@epa.gov

Gordon Lancaster, Transportation Operations Specialist, U.S. General Services Administration, (303) 236-7599, gordon.lancaster@gsa.gov

New York

Incentives: Funding is available for alternative fuel product development through the New York State Energy Research and Development Authority's (NYSERDA) Transportation Research Program. In addition, schools, municipalities, and transit agencies can get funding to purchase alternative fuel buses and supporting infrastructure. For more information, contact Patrick Bolton, Project Manager, NYSERDA, (518) 862-1090 x 3322, ppb@nyserda.org, or visit the agency's web site at http://www.nyserda.org/programs/transportation/afv.asp.

Regulations: Motor fuel franchise dealers are allowed to buy alternative fuels from suppliers who are not a franchise distributor regardless of the contract between the dealer and its existing franchise distributor. At least 2% of the state fleet's fuel use must be biodiesel; by 2012, at least 10% of the state's fleet must use biodiesel. For additional information, contact Clean Fueled Vehicles Council, New York State Office of General Services, (518) 473-6594, nys.clean.fuel@ogs.state.ny.us.

Contacts:

Linda Hardie, Clean Cities Coordinator, Clean Communities of Western New York, Inc., (716) 634-1038, ccofwny.lmh@verizon.net

Andria Adler, Clean Cities Coordinator, Greater Long Island Clean Cities Coalition, Inc., (631) 969-3700 x45, aadler@lift.org

Deborah Stacey, Clean Cities Coordinator, Capital District (Albany) Clean Communities, (518) 458-2161, dstacey@cdtcmpo.org

Paul Heaney, Clean Cities Coordinator, Genesee Region Clean Communities, (585) 624-8182, pheaney@worldnet.att.net

Joseph Barry, Clean Cities Facilitator, Clean Communities of Central New York/Onondaga Community, (315) 498-2548, barryj@sunyocc.edu

Lou Calcagno, Clean Cities Coordinator, New York City Clean Cities Coalition, (212) 487-6820, lcalcagno@dot.nyc.gov

Mike Scarpino, Project Manager, U.S. Department of Energy, National Energy Technology Laboratory, (412) 386-4726, michael.scarpino@netl.doe.gov

Patrick Bolton, Project Manager, NYSERDA, (518) 862-1090 x 3322, ppb@nyserda.org

Clean Fueled Vehicles Council, New York State Office of General Services, (518) 473-6594, nys.clean.fuel@ogs.state.ny.us

John Shipman, Chief Automotive Engineer, Con Edison, CNG Fleet Information, (718) 204-4009, shipmanj@coned.com

Ronald J. Gulmi, Lead Account Executive, Keyspan Energy Delivery, (516) 545-5164, rgulmi@keyspanenergy.com

Kerry-Jane King, Supervisor, Electric Transportation, New York Power Authority, (914) 390-8207, kerry-jane.king@nypa.gov

Joe Darling, Director of Equipment Management Division, New York Department of Transportation, (518) 457-2875, jdarling@dot.state.ny.us

Tristan Gillespie, Pollution Prevention Coordinator, Region 2, U.S. Environmental Protection Agency, (212) 637-3753, gillespie.tristan@epa.gov

Reema Loutan, Environmental Engineer, Region 2, U.S. Environmental Protection Agency, (212) 637-3760, loutan.reema@epa.gov

North Carolina

Incentives: A tax credit is available to biodiesel providers who make a minimum of 100,000 gallons of fuel annually. Biodiesel processors also can get a tax credit for their construction, facilities, and equipment costs. Furthermore, those who construct, buy, or rent biodiesel-related property can get a property tax credit. Refueling facilities that sell biodiesel can get a tax credit related to their infrastructure costs.

New biodiesel distributors and dealers of B100 can get a rebate through the North Carolina Soybean Producers Association. For more information, call the Association at (919) 839-5700 or visit its web site at http://www.ncsoy.org/

Regulations: State-owned vehicle fleets that comprise more than ten vehicles intended for highway use must reduce their petroleum use by 20% by January 1, 2010. This can be achieved, in part, through the use of biodiesel.

Contacts:

Tobin Freid, Clean Cities Coordinator, Triangle Clean Cities Coalition, (919) 558-9400, tcc@tjcog.org

Jason Wager, Clean Cities Coordinator, Centralina Clean Fuels Coalition, (704) 348-2707, jwager@centralina.org

Sarah Niess, Clean Cities Co-Coordinator, Centralina Clean Fuels Coalition, (704) 348-2719, sniess@centralina.org

Bill Eaker, Clean Cities Coordinator, Land of Sky Regional Council, Clean Vehicles Coalition (Not Yet Designated), (828) 251-6622 x118, bill@landofsky.org

Steven Richardson, Project Manager, U.S. Department of Energy, National Energy Technology Laboratory, (304) 285-4185, steven.richardson@netl.doe.gov

Heather Hildebrandt, Division of Air Quality, Mobile Source Compliance Branch, Department of Environment and Natural Resources, (919) 733-1498, heather.hildebrandt@ncmail.net

Cynthia Moseley, Alternative Fuels Program Manager, State Energy Office, (919) 733-1896, cynthia.moseley@ncmail.net

Anne Tazewell, Alternative Fuels Program Manager, North Carolina Solar Center, North Carolina State University, (919) 513-7831, anne_tazewell@ncsu.edu

Transportation Services, North Carolina Department of Public Instruction, (919) 807-3570

Greg Johnson, Manager, Business Support Services, Piedmont Natural Gas Company, (704) 731-4392, greg.johnson@piedmontng.com

Lydia McIntyre, Transportation Planning Engineer, Greensboro Department of Transportation, (336) 373-3117, lydia.mcintyre@greensboro-nc.gov

John M. Burns, Jr., Staff Engineer, Office of Chief Engineer, Division of Highways, North Carolina Department of Transportation, (919) 715-5658, jburns@dot.state.nc.us

North Carolina Soybean Producers Association, (919) 839-5700

Dale Aspy, Environmental Engineer, Region 4 Air Planning Branch, U.S. Environmental Protection Agency, (404) 562-9041, aspy.dale@epa.gov

Alan Powell, Environmental Engineer, Region 4 Air Planning Branch, U.S. Environmental Protection Agency, (404) 562-9045, powell.alan@epa.gov

Wes Allen, Transportation Specialist, Southeast Region, U.S. General Services Administration, (404) 331-3052, james.allen@gsa.gov

North Dakota

Incentives: A tax credit is available on the sale of biodiesel equipment. There is a biodiesel income tax credit as well. Additionally, there is a biodiesel production equipment tax credit, which is a corporate income tax credit. Furthermore, the Biofuels Partnership in Assisting Community Expansion (PACE) Loan Program provides financial assistance for several biodiesel activities. For information, contact Bob Humann, Senior Vice President of Lending, Bank of North Dakota, at (701) 328-5703 or (800) 472-2166 x5703, bhumann@nd.gov or visit the web site at http://www.banknd.com/ls/ls_commercial5.jsp.

Regulations: Retailers must ensure that alternative fuels are labeled. Biofuels are given a special excise tax rate. The sale of biodiesel equipment to a facility that sells biodiesel blends of B2 or higher is exempt from sales tax.

Contacts:

Dan Daly, Clean Cities Coordinator, Red River Valley/Winnipeg Manitoba Clean Cities Coalition, (701) 777-2822, ddaly@undeerc.org

Ernie Oakes, Project Manager, U.S. Department of Energy, Golden Field Office, (303) 275-4817, ernie.oakes@go.doe.gov

Joe Murphy, State Energy Program Administrator, North Dakota Department of Commerce, (701) 328-2697, jmurphy@nd.gov

Bob Humann, Senior Vice President of Lending, Bank of North Dakota, (701) 328-5703 or (800) 472-2166 x5703, bhumann@nd.gov

Gordon Lancaster, Transportation Operations Specialist, U.S. General Services Administration, (303) 236-7599, gordon.lancaster@gsa.gov

Ohio

Incentives: None known

Regulations: State agencies will receive credits for biodiesel use in new state diesel motor vehicles. Additionally, the state's Department of Transportation must use a minimum amount of biodiesel annually.

Contacts:

John McGovern, Clean Cities Coordinator, Earth Day Coalition's Clean Fuels Program, (216) 281-6468 x223, jmcgovern@earthdaycoalition.org

Sam Spofforth, Clean Cities Coordinator, Clean Fuels Ohio, (614) 292-5435, sam@cleanfuelsohio.org

Jim Zuber, Assistant Manager/Engineering Administrator, Ohio Department of Development's Office of Energy Efficiency, (614) 387-2731, jzuber@odod.state.oh.us

Mike Scarpino, Project Manager, U.S. Department of Energy, National Energy Technology Laboratory, (412) 386-4726, michael.scarpino@netl.doe.gov

Dale Palmer, Administrator, Ohio Office of Fleet Management, (614) 995-5296, dale.palmer@das.state.oh.us

Preston Boone, Energy Outreach Analyst, Ohio Department of Development, Office of Energy Efficiency, (641) 644-8864 or (866) 728-6749, plboone@odod.state.oh.us

Anita Field, Executive Secretary, Ohio Propane Gas Association, (614) 221-1900 ext. 215, opga@assnoffices.com

Mike McKay, Manager, Technology Business Assistance Office / Team Leader, Ohio Fuel Cell Initiative, Ohio Department of Development, Technology Division, (614) 644-9159, mmckay@odod.state.oh.us

Matthew Lindsay, Manager, Environmental Planning, Miami Valley Regional Planning Commission, (937) 223-6323, mlindsay@mvrpc.org

Mark Gnatowski, Administrator, Ohio Department of Transportation, Office of Equipment Management, (614) 351-2828, mgnatows@dot.state.oh.us

Scott Benson, Transportation Specialist, Great Lakes Region, U.S. General Services Administration, (312) 886-8682, scott.benson@gsa.gov

Oklahoma

Incentives: A tax credit is available for biodiesel production facilities.

Regulations: All school districts must consider operating their vehicles on alternative fuels whenever feasible.

Contacts:

Yvonne Anderson, Clean Cities Coordinator, Central Oklahoma Clean Cities Coalition, (405) 234-2264 x275, yanderson@acogok.org

Nancy Graham, Clean Cities Program Manager, Tulsa Area Clean Cities Coalition, (918) 584-7526, ngraham@incog.org

Neil Kirschner, Project Manager, U.S. Department of Energy, National Energy Technology Laboratory, (412) 386-5793, neil. kirschner@netl.doe.gov

Vaughn Clark, Director, Office of Community Development, Oklahoma Department of Commerce, (405) 815-5370, vaughn_ clark@odoc.state.ok.us

Clayton Robinson, SEP Projects/Alternative Fuels Loan Program Officer, Oklahoma Department of Commerce, State Energy Office, (405) 815-5249, clayton_robinson@odoc.state.ok.us

Richard Bailey, CNG and NGV Service, Maintenance and ManufacturingSupervisor, Oklahoma Natural Gas Company, (918) 640-1591, rbailey@ong.com

Gary Marchbanks, Manager of Government Accounts, Oklahoma Gas and Electric, Electric Services, (405) 553-8188, marchbgj@oge.com

Gordon Lancaster, Transportation Operations Specialist, U.S. General Services Administration, (303) 236-7599, gordon.lancaster@gsa.gov

Sandra Rennie, Mobile Source Team Leader, Region 6, U.S. Environmental Protection Agency, (214) 665-7367, rennie.sandra@epa.gov

Oregon

Incentives: Oregon's Department of Energy provides loans for the development of a renewable fuels refueling infrastructure. For information, contact Jeff Kato, Loan Manager, Oregon Department of Energy, (503) 373-7981, jeff.s.keto@state.or.us, or visit the agency's web site at http://egov.oregon.gov/energy/loans.

Portland's Office of Sustainable Development offers two grants for biofuels. For information, contact the City of Portland Office of Sustainable Development, (503) 823-7222, or visit the web site at http://www.portlandonline.com/OSD.

Regulations: All of the City of Portland's diesel vehicles must use a biodiesel blend of B20 or higher. Meanwhile, all petroleum diesel fuel sold within Portland's city limits must contain a minimum of 5% biodiesel; this number will increase to 10% by July 1, 2010.

Contacts:

Steve Vincent, Clean Cities Coordinator, Rogue Valley Clean Cities Coalition, (541) 858-4773, steve.vincent@avistacorp.com

Rick Wallace, Clean Cities Coordinator, Columbia Willamette Clean Cities Coalition, Inc., (503) 378-3265, rick.wallace@state.or.us

Ernie Oakes, Project Manager, U.S. Department of Energy, Golden Field Office, (303) 275-4817, ernie.oakes@go.doe.gov

Deby Davis, RETC Program Specialist, Oregon State Energy Office, (503) 378-4040 x291, deby.s.davis@state.or.us

Jeff Keto, Loan Manager, Oregon Department of Energy, (503) 373-7981, jeff.s.keto@state.or.us

Justin Klure, Senior Policy Analyst, Oregon State Energy Office, (503) 373-1581, justin.klure@state.or.us

Gordon Griffin, Program Manager, Lane Regional Air Protection Agency, (541) 736-1056, gordon@lrapa.org

Sharon Banks, Founder, CEO, Cascade Sierra Solutions, (541) 302-0900, sharon@cascadesierrasolutions.org

City of Portland Office of Sustainable Development, (503) 823-7222

Chris Galati, Director, Conservation and Technology, NW Natural Gas, (503) 721-2472, cfg@nwnatural.com

Brian Boothe, Fleet Service Representative (Western Oregon), U.S. General Services Administration, Pacific Cascade Fleet Management Center, (360) 696-7541, brian.boothe@gsa.gov

Julie Shain, Fleet Manager (Eastern Oregon), U.S. General Services Administration, (208) 321-9150, julie.shain@gsa.gov

Pennsylvania

Incentives: Alternative fuel grants are available to small businesses interested in adopting energy-efficient or pollution prevention measures or equipment. For information, contact David Barnes, Program Manager, Pennsylvania Department of Environmental Protection, Office of the Small Business Ombudsman, (717) 772-5160, dbarnes@state.pa.us, or contact the agency's web site at http://www.depweb.state.pa.us.

Regulations: The PennSecurity Fuels Initiative requires the use of biodiesel and other potentially renewable fuels. It also is developing incentives for Pennsylvania farmers who grow renewable energy feedstock. For additional information, visit http://www.depweb.state.pa.us.

Contacts:

Rick Price, Clean Cities Co-Coordinator, Pittsburgh Region Clean Cities Coalition, (412) 386-6196, price@netl.doe.gov

Nathaniel Doyno, Clean Cities Co-Coordinator, Pittsburgh Region Clean Cities Coalition, (412) 418-4594, nathaniel@steelcitybiofuels.org

Brinda Shetty, Clean Cities Coordinator, Greater Philadelphia Clean Cities Program, (215) 413-2122, coordinator@phillycleancities.org

Kay Milewski, Project Manager, U.S. Department of Energy, National Energy Technology Laboratory, (304) 285-4535, kay.milewski@netl.doe.gov

Susan Summers, Alternative Fuels Program Manager, Pennsylvania Department of Environmental Protection, Office of Energy and Technology Deployment, (717) 783-9242, susummers@state.pa.us

David Barnes, Program Manager, Pennsylvania Department of Environmental Protection, Office of the Small Business Ombudsman, (717) 772-5160, dbarnes@state.pa.us

Gene DelVecchio, Program Manager, Pennsylvania Department of Environmental Protection, (717) 772-8951, gdelvecchi@state.pa.us

Pieter Ouwerkerk, Senior Engineer of Gas Engineering, PECO Energy Company, (215) 841-5220, pieter.ouwerkerk@exeloncorp.com

Barry Wentzel, Manager, I and C Marketing, UGI Utilities, (610) 796-3548, bwentzel@ugi.com

Mack Godfrey, Columbia Gas of Pennsylvania, Inc., (724) 416-6324, mgodfre@nisource.com

Michael Golembiewski, Transportation Modeler, Berks County Planning Commission, (610) 478-6300, planning@countyofberks.com

Reynold L. Sanner, Western Zone Manager, Region 3, U.S. General Services Administration, (724) 693-2400 x4, reynold.sanner@gsa.gov

Rhode Island

Incentives: Taxpayers can use the Alternative Fueled Vehicle and Filling Station Tax Credit toward a portion of the labor, capital, and equipment costs associated with an alternative fuel refueling station. Additionally, a fuel tax deduction is available to corporations that sell alternative fuels. Finally, biodiesel made from organically grown feedstock is not subject to motor fuel tax.

Regulations: None known

Contacts:

Jennifer Cole, Clean Cities Coordinator, Ocean State Inc. Clean Cities Coalition, (401) 351-6440 x15, cleancitiesri@gmail.com

Mike Scarpino, Project Manager, U.S. Department of Energy, National Energy Technology Laboratory, (412) 386-4726, michael. scarpino@netl.doe.gov

Tim Howe, Principal Planner, Rhode Island State Energy Office, (401) 222-3370 x104, timh@gw.doa.state.ri.us

Janice McClanaghan, Chief of Energy and Community Services, Rhode Island State Energy Office, (401) 222-3370 x109, janicem@ gw.doa.state.ri.us

Office of Air Resources, Rhode Island Department of Environmental Management, (401) 222-2808

Frank Stevenson, Supervising Air Quality Specialist, Department of Environmental Management, (401) 222-4700 x7021, frank.stevenson@dem.ri.gov

Robert Judge, Environmental Engineer, Region 1, U.S. Environmental Protection Agency, (617) 918-1045, judge.robert@epa. gov

Andrew E. Motter, Community Planner, U.S. Department of Transportation, Federal Transit Administration, Region 1, (617) 494-3560, andy.motter@fta.dot.gov

South Carolina

Incentives: For every gallon of B20 sold, biodiesel retailers can get a 5-cent-per-gallon payment. A biodiesel production tax credit is available and varies in amount depending on the type of feedstock used. Finally, a tax credit is available for biodiesel facilities.

Regulations: Although alternative fuels and fuel blends are exempt from the state sales and use tax, they are subject to a state fuels tax.

Contacts:

Wendy Bell, Palmetto State Clean Fuels Coalition, Clean Cities Coordinator/Senior Planner, (803) 327-9041, (803) 327-1912, wbell@catawbacog.org

Steven Richardson, U.S. Department of Energy, National Energy Technology Laboratory, Project Manager, (304) 285-4185, steven.richardson@netl.doe.gov

John White, South Carolina Department of Transportation, Director of Supply and Equipment, (803) 737-6675, (803) 737-6680, whitejf@scdot.org

Greg Johnson, Piedmont Natural Gas Company, Manager of Business Support Services, (704) 731-4392, (704) 364-8320, greg.johnson@piedmontng.com

Dale Aspy, U.S. Environmental Protection Agency, Environmental Engineer, Region 4 Air Planning Branch, (404) 562-9041, (404) 562-9019, aspy.dale@epa.gov

Wes Allen, U.S. General Services Administration, Transportation Specialist, Southeast Region, (404) 331-3052, (404) 331-1879, james.allen@gsa.gov

South Dakota

Incentives: A biodiesel production facility tax refund is available for projects that cost more than $4.5 million.

Regulations: All state Department of Transportation diesel vehicles must use a minimum blend of B2. Biodiesel is taxed at a lower rate than are other motor fuels.

Contacts:

Ernie Oakes, Project Manager, U.S. Department of Energy, Golden Field Office, (303) 275-4817, ernie.oakes@go.doe.gov

Chuck Fergen, Financial Analyst, South Dakota Department of Transportation, (605) 773-4114, chuck.fergen@state.sd.us

David Knigge, Assistant Finance Director, South Dakota Department of Transportation, (605) 773-4555, david.knigge@state.sd.us

Gordon Lancaster, Transportation Operations Specialist, U.S. General Services Administration, (303) 236-7599, gordon.lancaster@gsa.gov

Tennessee

Incentives: The state's Department of Transportation is allowed to develop partnerships with farmer cooperatives, transportation fuel providers, and others in order to install biodiesel refueling facilities. Furthermore, grants are available to those developing biodiesel infrastructure. For information, contact Terry Ellis, Program Manager, Tennessee State Energy Office, Department of Economic and Community Development, Energy Division, (800) 342-1340 or (615) 741-2994, terry.ellis@state.tn.us, or visit the web site at http://www.state.tn.us/ecd/energy_biodiesel.htm.

Regulations: Tennessee's Department of Agriculture is allowed to inspect and test biofuels.

Contacts:

Jonathan Overly, Executive Director, East Tennessee Clean Fuels Coalition, (865) 974-3625, jgoverly@utk.edu

Dave Pelton, Clean Cities Coordinator, Clean Cities of Middle Tennessee, (615) 482-4849, davepelton@att.net

Andrew Couch, Clean Cities Coordinator, West Tennessee Clean Cities Coalition (Not Yet Designated), (901) 378-2497, andrew@ wtccc.com

Steven Richardson, Project Manager, U.S. Department of Energy, National Energy Technology Laboratory, (304) 285-4185, steven. richardson@netl.doe.gov

Terry Ellis, Program Manager, Tennessee State Energy Office, Department of Economic and Community Development, Energy Division, (800) 342-1340 or (615) 741-2994, terry.ellis@state.tn.us

Walter C. Miller, Energy Services Consultant, Atmos Energy, (817) 303-2903, walter.c.miller@atmosenergy.com

Parks Wells, Tennessee Soybean Promotion Council, (731) 668-2850, pwells@tnsoybeans.org

Dale Aspy, Environmental Engineer, Region 4 Air Planning Branch, U.S. Environmental Protection Agency, (404) 562-9041, aspy.dale@epa.gov

Wes Allen, Transportation Specialist, Southeast Region, U.S. General Services Administration, (404) 331-3052, james.allen@gsa.gov

Texas

Incentives: Alternative fuel grants are available through the Texas Emissions Reduction Plan (TERP). For information, contact Dr. Rudy Smaling, PhD, NTRD Program Director, Senior Research Scientist, Houston Advanced Research Center, (281) 364-4036, rsmaling@harc.edu, or visit the web site at http://www.harc.edu.

Regulations: Biodiesel is exempt from the petroleum diesel fuel tax.

Contacts:

Andy Hudgins, Clean Cities Coordinator, Alamo Area Clean Cities Coalition, (210) 362-5200

Stacy Neef, Clean Cities Coordinator, Central Texas Clean Cities Coalition, (512) 482-5343, stacy.neef@austinenergy.com

Mindy Mize, Clean Cities Coordinator, Dallas/Fort Worth Clean Cities Coalition, (817) 608-2346, mmize@nctcog.org

Stephanie Lee, Clean Cities Coordinator, Greater Houston Regional Clean Cities Coalition, (713) 993-4581, stephanie.lee@h-gac.com

Rick McKnight, Clean Cities Coordinator, East Texas Clean Cities Coalition, (903) 984-8641, rick.mcknight@etcog.org

Miguel Conchas, Acting Clean Cities Coordinator, Laredo Clean Cities Coalition, (956) 722-9895, conchas@laredochamber.com

Bob Dickinson, Clean Cities Co-Coordinator, South East Texas Clean Cities Coalition, (409) 899-8444 x251, bdickinson@setrpc.org

Dawn Martinez, Clean Cities Co-Coordinator, South East Texas Clean Cities Coalition, (409) 899-8444 x253, dmartinez@setrpc.org

Neil Kirschner, Project Manager, U.S. Department of Energy, National Energy Technology Laboratory, (412) 386-5793, neil.kirschner@netl.doe.gov

Alternative Fuels Research and Education Division (AFRED), Railroad Commission of Texas, (512) 463-7110

Heather Ball, AFRED Assistant Director, Marketing and Public Education, Railroad Commission of Texas, (512) 463-7359, heather.ball@rrc.state.tx.us

Dan Kelly, AFRED Director, Railroad Commission of Texas, (512) 463-7291, dan.kelly@rrc.state.tx.us

Eileen Latham, AFRED Rebate Coordinator, Railroad Commission of Texas, (512) 475-2911 or (800) 64-CLEAR, eileen.latham@rrc.state.tx.us

Franz Hofmann, AFRED Lead Automotive Instructor, Railroad Commission of Texas, (512) 463-8501, franz.hofmann@rrc.state.tx.us

Soll Sussman, Alternative Fuels Program Coordinator, Texas General Land Office, Energy Resources, (512) 463-5039, soll.sussman@glo.state.tx.us

Steve Dayton, Team Leader of Grant Contract Development Team, Texas Commission on Environmental Quality, (512) 239-6824, sdayton@tceq.state.tx.us

Dr. Rudy Smaling, PhD, NTRD Program Director, Senior Research Scientist, Houston Advanced Research Center, (281) 364-4036, rsmaling@harc.edu

Mary-Jo Rowan, Program Manager, Texas State Energy Conservation Office, Texas Comptroller of Public Accounts, (512) 463-2637, mary-jo.rowan@cpa.state.tx.us

Teri Green, Conservation Program Manager, Texas Gas Service, (512) 465-1109, tgreen@txgas.com

Walter C. Miller, Energy Services Consultant, Atmos Energy, (817) 303-2903, walter.c.miller@atmosenergy.com

Don Lewis, Fleet Manager, Texas State Department of Transportation, General Services Division, (512) 374-5471, dlewis1@dot.state.tx.us

Sandra Rennie, Mobile Source Team Leader, Region 6, U.S. Environmental Protection Agency, (214) 665-7367, rennie.sandra@epa.gov

Gordon Lancaster, Transportation Operations Specialist, U.S. General Services Administration, (303) 236-7599, gordon.lancaster@gsa.gov

Utah

Incentives: None known
Regulations: None known

Contacts:

Robin Erickson, Director, Utah Clean Cities Coalition, (801) 535-7736, Robin.Erickson@slcgov.com

Ernie Oakes, Project Manager, U.S. Department of Energy, Golden Field Office, (303) 275-4817, ernie.oakes@go.doe.gov

Ran Macdonald, Environmental Engineer, Utah Department of Environmental Quality, Division of Air Quality, (801) 536-4071, rmacdonald@utah.gov

Glade Sowards, Energy Program Coordinator, Utah Division of Air Quality, (801) 536-4020, gladesowards@utah.gov

Maylene Roach, Accountant, Salt Lake City Department of Airports, (801) 575-2048, maylene.roach@slcgov.com

Dan Bergenthal, Transportation Engineer, Salt Lake City Transportation Division, (801) 535-6630

Gordon Larsen, Natural Gas Vehicle Supervisor, Questar Gas, (801) 324-5987, gordon.larsen@questar.com

Jim Grambihler, Natural Gas Vehicle Operations, Questar Gas, (801) 324-5119, jim.grambihler@questar.com

Doug Anderson, Project Manger, Research, Utah Department of Transportation, (801) 965-4377, dianderson@utah.gov

Utah State Tax Commission Motor Vehicle Division, (800) DMV-UTAH or (801) 297-7780, dmv@utah.gov

Gordon Lancaster, Transportation Operations Specialist, U.S. General Services Administration, (303) 236-7599, gordon.lancaster@gsa.gov

Vermont

Incentives: Vermont businesses involved in alternative energy technology might be eligible for a tax credit.

Regulations: None known

Contacts:

Karen Glitman, Program Coordinator, UVM Transportation Center/Clean Cities Coordinator, Vermont Clean Cities Coalition, (802) 656-8868, karen.glitman@uvm.edu

Mike Scarpino, Project Manager, U.S. Department of Energy, National Energy Technology Laboratory, (617) 565-9716, michael.scarpino@netl.doe.gov

Debra Baslow, Buildings Engineer, Department of Buildings and General Services, (802) 828-0377, debra.baslow@state.vt.us

James (J.J.) Mullowney, Manager, Technical Services, Vermont Gas Systems, (802) 863-4511 x339, jmullowney@vermontgas.com

Gina Campoli, Environmental Policy Manager, Vermont Agency of Transportation, Policy and Planning Division, (802) 828-5756, gina.campoli@state.vt.us

Robert Judge, Environmental Engineer, Region 1, U.S. Environmental Protection Agency, (617) 918-1045, judge.robert@epa.gov

Andrew E. Motter, Community Planner, U.S. Department of Transportation, Federal Transit Administration, Region 1, (617) 494-3560, andy.motter@dot.gov

Virginia

Incentives: Grants are available to large-scale (annual minimum two million gallons) biodiesel producers through the Biofuels Production Fund.

Regulations: All state diesel vehicles must use a B20 biodiesel blend whenever feasible.

Contacts:

Chelsea Jenkins, Clean Cities Coordinator, Hampton Roads Clean Cities Coalition, (757) 256-8528 or (888) 276-3320, cjenkins@hrccc.org

George Nichols, Washington Metropolitan Clean Cities Coordinator, Metropolitan Washington Council of Governments, (202) 962-3355, gnichols@mwcog.org

Kay Milewski, Project Manager, U.S. Department of Energy, National Energy Technology Laboratory, (304) 285-4535, kay.milewski@netl.doe.gov

John Warren, Division Director of Energy, Virginia Department of Mines, Minerals and Energy, (804) 692-3216, john.warren@dmme.virginia.gov

Jimmy Conroy, Fleet Supervisor, Virginia Natural Gas, (757) 466-5506, jconroy@aglresources.com

Richard Rasmussen, Director of Small Business Assistance (Environmental Compliance Assistance Loan), Virginia Department of Environmental Quality, (804) 698-4394, rgrasmussen@deq.virginia.gov

John Carlock, Deputy Executive Director for Physical Planning, Hampton Roads Planning District Commission, (757) 420-8300, jcarlock@hrpdc.org

Don Unmussig, Director of Fleet Management Services, Virginia Department of General Services, (804) 367-6525, donald.unmussig@dgs.virginia.gov

Walter C. Miller, Energy Services Consultant, Atmos Energy, (817) 303-2903, walter.c.miller@atmosenergy.com

Virginia Department of Motor Vehicles, (866) 368-5463 or (800) 435-5137

Washington

Incentives: A tax deduction is available for the sale or distribution of biodiesel. In addition, biofuels retail tax and production tax exemptions are available. An alternative fuel grant and loan program exists for renewable and alternative fuel infrastructure, facilities, and technologies; it also funds research and development of alternative fuel markets.

Regulations: Starting on November 30, 2008, all petroleum diesel sold in Washington must contain a minimum of 2% biodiesel. Starting on June 1, 2009, all state agencies' diesel vehicles must use a minimum of B20. Starting on June 1, 2015, 100% of the fuel used by state and local government agencies must be from biofuels or electricity. Additionally, certain public authorities and conservation districts are allowed to enter into biodiesel production contracts. Finally, biodiesel blends – but not B100 – are subject to Underground Storage Tank regulations. Anyone who owns a diesel tank that will now hold B100 needs to communicate that change to the Washington State Department of Ecology and have a certified site assessor perform a site assessment. For additional information, contact Gail Jaskar ((360) 407-7225 or (800) 826-7716 or gjas461@ecy.wa.gov) or Mike Blum ((360) 407-6913 or mblu461@ecy.wa.gov) of the Washington State Department of Ecology, http://www.ecy.wa.gov/pubs/0309103.pdf.

Contacts:

Mark H. Brady, Clean Cities Coordinator, Puget Sound Clean Cities Coalition, (206) 684-0935, bradymh@seattle.gov
Ernie Oakes, Project Manager, U.S. Department of Energy, Golden Field Office, (303) 275-4817, ernie.oakes@go.doe.gov
Chuck Dougherty, Program Manager for Alternative Fuel Vehicles, Puget Sound Energy, (253) 476-6202, chuck.dougherty@pse.com

Kim Lyons, , Washington State Energy Office, Washington State University Energy Program, (360) 956-2083, lyonsk@energy.wsu.edu

Dean Lookingbill, Transportation Director, Southwest Washington Regional Transportation Council, (360) 397-6067 x5208, dean.lookingbill@rtc.wa.gov

Kelly McGourty, Principal Planner, Puget Sound Regional Council, (206) 464-7892, kmcgourty@psrc.org

Mia Waters, Air Quality, Acoustics and Energy Programs Manager, Washington State Department of Transportation, (206) 440-4541, watersy@wsdot.wa.gov

Gail Jaskar, Washington State Department of Ecology, (360) 407-7225 or (800) 826-7716, gjas461@ecy.wa.gov

Mike Blum, Underground Storage Tank Coordinator, Washington State Department of Ecology, (360) 407-6913, mblu461@ecy.wa.gov

Jill Simmons, Seattle's Office of Sustainability and Environment, (206) 684-9261, jill.simmons@seattle.gov

Julie Shain, Fleet Manager, U.S. General Services Administration, (208) 321-9150, julie.shain@gsa.gov

West Virginia

Incentives: A portion of the cost for alternative fuel use for a West Virginia school bus can be reimbursed; a plan detailing the proposed use of the alternative fuel must be submitted to the state's Department of Education.

Regulations: No political subdivisions can offer incentives or subsidies for alternative fuels production.

Contacts:

Kelly Bragg, Coordinator, West Virginia Clean State Program, (304) 558-2234 or (800) 982-3386, kbragg@wvdo.org

Kay Milewski, Project Manager, U.S. Department of Energy, National Energy Technology Laboratory, (304) 285-4535, kay.milewski@netl.doe.gov

Al Ebron, Executive Director, West Virginia University, National Alternative Fuels Training Consortium, (304) 293-7882, al.ebron@mail.wvu.edu

Reynold L. Sanner, Western Zone Manager, Region 3, U.S. General Services Administration, (724) 693-2400 x4, reynold.sanner@gsa.gov

Wisconsin

Incentives: The Wisconsin Department of Public Instruction (DPI) is allowed to help school districts to pay for school bus biodiesel costs. Anyone who uses at least 100 gallons of an alternative fuel in a taxi can get a tax refund.

Regulations: The Wisconsin Department of Administration must reduce its use of petroleum diesel by 10% in 2010 and by 25% in 2015. Biodiesel sellers must follow the state's labeling requirements for biodiesel fuel.

Contacts:

Francis Vogel, Clean Cities Coordinator, Wisconsin Clean Cities Southeast Area, Inc., (414) 221-4958, francis.vogel@we-energies.com

Mike Scarpino, Project Manager, U.S. Department of Energy, National Energy Technology Laboratory, (412) 386-4726, michael.scarpino@netl.doe.gov

Muhammed Islam, Vehicle Emissions Engineer, Wisconsin Department of Natural Resources, Clean Fuel Fleet Program, (608) 264-9219, muhammed.islam@wisconsin.gov

Jessica Lawent, Air Quality Program Specialist, Wisconsin Department of Natural Resources, Clean Fuel Fleet Program, (414) 263-8653, jessica.lawent@wisconsin.gov

Jean Beckwith, Bureau of Entrepreneurship, Wisconsin Department of Commerce, (608) 261-2517, jean.beckwith@wisconsin.gov

Bob Reagan, Project Manager, We Energies, (414) 221-2284, bob.reagan@we-energies.com

Maria Redmond, Alternative Fuels Policy Analyst, State of Wisconsin, Department of Agriculture, Trade and Consumer Protection, (608) 224-4607, maria.redmond@wisconsin.gov

Excise Tax Section, Wisconsin Department of Revenue, (608) 266-3223 or (608) 266-0064, excise@dor.state.wi.us

Ken Neusen, Director, University of Wisconsin-Milwaukee, Center for Alternative Fuels, Wisconsin Alternative Fuels Task Force, (414) 229-4272, neusen@uwm.edu

Scott Benson, Transportation Specialist, Great Lakes Region, U.S. General Services Administration, (312) 886-8682, scott.benson@gsa.gov

Wyoming

Incentives: None known
Regulations: None known

Contacts:

Sandy Shuptrine, Clean Cities Coordinator/Executive Director, Greater Yellowstone/ Teton Clean Energy Coalition, 307-733-6371, sandyshuptrine@wyom.net

Ernie Oakes, Project Manager, U.S. Department of Energy, Golden Field Office, (303) 275-4817, ernie.oakes@go.doe.gov

Tom Fuller, Manager, State Energy Program, Wyoming Energy Office, (307) 777-2804, tom.fuller@wybusiness.org

Don Bainter, Energy Services Representative, Cheyenne Light, Fuel and Power, (307) 778-2133, dbainter@cheyennelight.com

Bill Gray, Energy Services Representative, Cheyenne Light, Fuel and Power, (307) 778-2145, bgray@cheyennelight.com

Gordon Larsen, Natural Gas Vehicle Supervisor, Questar Gas, (801) 324-5987, gordon.larsen@questar.com

Jim Grambihler, Natural Gas Vehicle Operations, Questar Gas, (801) 324-5119, jim.grambihler@questar.com

Rich Douglass, Local Government Coordinator, Wyoming Department of Transportation, (307) 777-4384, rich.douglass@dot.state.wy.us

Gordon Lancaster, Transportation Operations Specialist, U.S. General Services Administration, (303) 236-7599, gordon.lancaster@gsa.gov

Case Study: Unidentified Delinquent

Wanting to do the right thing, a home brewer of biodiesel applied online to his state for an "alternative fuels road tax registration certificate" that he understood would allow him to pay a one-time, $50.00 per annum, pro-rated fee for the road use taxes owed on his homeproduced fuel. He downloaded the form and filled out the application and sent in his check only to receive a call two weeks later from the state road tax office saying that his application had been rejected because "biodiesel is not an alternative fuel." He listened in disbelief as a clerk told him that biodiesel is NOT an alternative fuel but that liquefied propane (LP), a non-renewable, fossil fuel is.

On further questioning of the director of the fuels road use tax department, the home brewer learned that, in his particular state, diesel fuel is defined in the state code as ANY fuel that powers a diesel engine. Thus, biodiesel or waste vegetable oil – or turkey droppings for that matter – would be considered taxable fuel; however, petroleum-based LP would be exempt from paying the road use taxes. This is a true testament to the power of lobbyists and particularly poignant in a state that prides itself in being a major agricultural producer.

This story ends with the home brewer receiving by mail his returned, un-cancelled check and a stack of forms with no instructions on HOW to fill them out. However, the packet did include a very comprehensible statement detailing the penalties and fines to be imposed if the home brewer should fail to file the enclosed forms monthly and in a prompt manner. At the time of this writing, the home brewer, having surrendered himself to local law enforcement officials and having confessed his sin, has yet to be arrested.

Chapter 7. Sustainable Biodiesel Certifications & Guidelines

This chapter explores the need for a biodiesel certification program that addresses environmental (including human) health and safety throughout the entire biodiesel life cycle. It reviews current efforts to create a sustainable biodiesel certification program. Then, it looks to other programs, such as Europe's environmental certification program for biodiesel, that could inform the development of such a program in the U.S.

This book has discussed the existing framework of biodiesel best practices, quality specifications, regulations, and permitting. If sustainability becomes a central tenet of the biofuels industry, then biodiesel and other biofuels hold great promise as we transition away from oil and toward truly renewable (e.g., solar) fuels.

It should be clear, however, that the existing framework of best practices, specifications, regulations, and permitting do not fully address sustainability. Although many of the regulations and best practices are geared toward environmental health and human safety, they generally do not take a long-range view of sustainability. They are doing very important things – such as keeping pollutants out of the air and water and reducing the chances that biodiesel producers will injure themselves and others or start fires. However, they are operating at the scale of individual producers, distributors, and others. That is important – but we also need a larger vision.

The Need to Differentiate
More Sustainable from Less Sustainable

Although biodiesel is a promising fuel for many countries, its production – particularly if done on a large scale and using feedstock from non-sustainable agricultural methods – can come at a high price. Millions of acres of forest around the world already have been replaced with biodiesel crops, resulting in the loss of ecological integrity, biological diversity, ecosystem resilience, natural capital, and evolutionary opportunities.

Yet, demand for biofuels is on the rise, and several countries have set goals for significant increases in biodiesel production. In order to meet those goals, companies are ramping up the production of feedstocks and fuel, often ignoring the fact that, despite cleaner emissions, the production of biodiesel can harm the environment that it is supposed to protect. One biodiesel producer whom we interviewed referred to his industry as "shameful."

The realization that today's biodiesel might not be the perfect replacement for petroleum products has caused some countries to reconsider their support of biodiesel as an alternative fuel. For example, National Express Group, one of the United Kingdom's largest transportation companies, has ended its biodiesel bus trials due to concern over the food-versus-fuel issue in which some are concerned that biodiesel crops will replace food crops. Similarly, the European Union's goal of meeting 10% of its transport fuel with biofuels by 2020 has been influenced by reports that much of the EU's agricultural lands would have to be converted to use for biofuels in order to meet that goal. Now, the EU is developing criteria for sustainable biofuels, and any biofuels that do not meet those criteria will not count toward the 10% goal.

After all, how can we ensure that biodiesel is sourced, produced, distributed, and used in such a way that it not only is environmentally benign but also supports local/regional economies and people? How do we ensure that the biodiesel industry grows sustainably and does not become an energy-intensive, centralized energy producer that is not sensitive to place, culture, or economy? How can consumers tell the difference between more sustainable biodiesel and less sustainable biodiesel?

A growing number of people – from biodiesel producers and governments to environmental groups and consumers – are asking questions like these. One way to address such questions is through the creation of an ecolabel. Ecolabels are logos that companies can use to identify their products or processes as environmentally preferable throughout the entire product/process life cycle. They generally result from some type of third-party audit and certification activity. (There are different types of certification process, such as the track and trace system, which would track the biodiesel throughout its life cycle.) There are several criteria that must be met before a product can be shown with the ecolabel (Townsend 2006). Ecolabels are somewhat straightforward in that a product/process either can display the ecolabel or it cannot.

There are many types of ecolabels used around the world. For instance, Germany's Blue Angel, introduced in 1977, is used to identify light bulbs, laundry detergent, and other products that are more environmentally friendly throughout their life cycles. Other ecolabels include the EU's Eco-label, Canada's Environmental Choice program, Japan's Eco Mark, France's NF *Environnement*, and the United States' Energy Star and Green Seal programs.

Another type of certification program is one based on a tiered system. In this type of system, one company might make several products that vary with regard to their degree of environmental preferability.

Let's suppose that Company A makes several products that are far greener than its competitor, Company B, makes. All of Company A's products deserve ecolabels – but some products are greener than others. Why not reward Company A for making all of those greener products *and* differentiate between those products to indicate those that are the greenest? That is exactly what the U.S. Green Building Council did through its LEED (Leadership in Energy and Environmental Design) program. This program was developed to identify more sustainable buildings. It is based on a point system; the more green features a building has, the more points it gets. Buildings that are rated the highest level are considered LEED Platinum. The next step down is LEED Gold, then LEED Silver, and so on. The tiered certification process is especially helpful when dealing with a complicated product, such as a building, which is made using thousands of products.

Case Study: Appalachian State University, Boone, NC

In the Spring of 2005, two undergraduate students at Appalachian State University (ASU) wrote a grant proposal for the EPA's P3 (People, Planet, and Prosperity) design competition for sustainability. The competition focused on innovation, marketability, and a multi-disciplinary focus. The ASU students, enrolled in the school's Appropriate Technology degree program, wanted to find a way to produce biodiesel sustainably. Its first year was funded, and the students won a second competition for an additional two years. Their project is called "Closing the Biodiesel Loop: Self Sustaining Community-Based Biodiesel Production."

Jeremy Ferrell, the project's coordinator, says, "When we started this, there really wasn't a good example of a self-sustaining, community-based processor that was replicable. We wanted to make something that was focused on using renewable energy for process inputs and on recycling the byproducts. Our goal was to address sustainable, community-scale processing focusing on closing the loops of waste and non-renewable resources."

The ASU group uses frying oil from the school's cafeteria as its biodiesel feedstock. What differentiates it from most other producers is the focus on using renewable energy to power the production process and recycling of waste. ASU uses photovoltaics to power the production process, a passive solar greenhouse to process the wastewater and for year-round production, and active solar thermal for process heating.

The students distill the methanol from the glycerin and test the methanol for quality. If its quality is good (containing <1% water), they mix it with new methanol and use it to make more biodiesel. The chemistry department uses gas chromatography to test for water in the methanol. Following methanol recovery, students use the glycerin for compost and soap making.

Are there any areas for improvement? Jeremy suggests, "There are better ways to make a better quality, more usable soap. I'd also like to see more work done on studying composting glycerin." He also believes that the ecological machine, based on John Todd's living machine, might be made more effective in treating the wastewater that results from the biodiesel production process. "There's a huge biochemical world that students can get involved in... resident time and flow and aeration. Whether you get an anaerobic tank or anoxic and that kind of thing." ASU also collects data on solar thermal operations, greenhouse performance, and PV output. Jeremy says, "I would like to see continued analysis of these data to hone the energy balance of our facility." On the people side of things, Jeremy hopes that the department will be able to pay for a graduate assistant to maintain the facility and coordinate outreach components like workshops and fuel distribution to off-road sectors.

As part of its commitment to sustainability, ASU also focuses on safety. It had to pass muster on the building codes, which meant that it needed an approved structure. ASU constructed a corrugated metal dome and insulated it. It had to consider issues such as egress, meaning the doors had to open a certain way, and the facility had to be handicapped accessible.

The fire codes allow only 60 gallons of methanol in the building at one time, which makes storing and recovering methanol a little tricky. The fire code also required them to put drywall over the insulation, which was challenging because the building is a corrugated metal dome. Since they are dealing with chemicals, they installed a sparkless fan for ventilation. During biodiesel processing, volatile vapors can vent from the processor tank as you displace air with methoxide. Sparkless motors are more common when you are pumping any kind of volatile material, like petroleum diesel or gas.

ASU overcame these issues and is in the process of purchasing new equipment and will continue using this opportunity to make biodiesel as a hands-on teaching tool for the students. ASU has tested and passed the total glycerin and free glycerin tests, which indicate the completeness of the reaction and how well the fuel is made. ASU also has had testing done by Foothills Bio-Energies in Lenoir, North Carolina, and feels confident that its fuel is ASTM quality. However, because the fuel is not made for commercial sale and goes into off-road use, ASU does not need to have it tested.

As standards develop for biodiesel, it will find more and more mainstream use. BQ9000 might evolve and become a good quality standard. Certainly, that kind of certification will happen, Jeremy says, and he would be interested in incorporating it into

ASU's biodiesel program. However, sustainability is about more than quality.

"What we try to do here at Appalachian is to show a model of how you can produce it sustainably. Despite all of the advances in technology, we have far more opportunity to make a difference with conservation." Biodiesel is very holistic, he explains, and it leads to other important discussions, like community development and the importance of living and walking in our communities. "We're trying to make cultural change. On the technical side, making biodiesel out of GMO crops is not really appealing, but it is a step out of the box using today's technology." Ultimately, however, sustainable biodiesel will require us to focus on micro-modal production using local materials and regional distribution. We will have to address issues of scale. He suggests, "Another way that biodiesel can become more sustainable is to give more incentives to renewable energy technology, like solar thermal and photovoltaics." These renewable energy technologies can be used to power the biodiesel manufacturing process.

As for the waste created by the biodiesel process, Jeremy says that the industry is likely to become more efficient at reusing its waste. First, they'll pull methanol off glycerin and sell it back to biodiesel plants. Then, the glycerin market, which had bottomed out due to the glycerin glut created by biodiesel, will find new life as companies find new uses for glycerin. "There's a company in Charlotte that has just started buying glycerin. There are some huge ADM and Cargill plants in Iowa that are just starting to buy glycerin. Propylene glycol, plasticizers – anything we can make from petroleum, we'll figure out how to make from glycerin," he says.

He also believes that people will be creative regarding biodiesel feedstock. Of course, algae is a promising feedstock but is technologically intense and requires growing it from industrial waste streams, like municipal wastewater or coal-fired power plants. Ultimately, Jeremy believes that biodiesel and other biofuels will be incorporated into national blends in concert with the petroleum industry. Biodiesel is a superior fuel for engine performance and longevity, and he thinks that there will be a lot of green marketing for blends like B5 and B20. As the green-shift continues, mainstream companies will feel increasingly confident and proud of commercializing homegrown renewable fuels. For more information about the collaborative biodiesel project at ASU, visit www.biodiesel.appstate.edu.

Sustainable Biodiesel Certification

Fortunately, there are groups that are working to address these and other issues that face the biodiesel community. Certification schemes would necessarily have to be easy to apply and flexible enough to take account of local conditions.

A few groups are working to develop an ecolabel for sustainable biodiesel. They include the Sustainable Biodiesel Alliance and the European Biodiesel Board. At the time of this writing, both organizations are developing their certification standards. A brief overview of each is provided below.

US: Sustainable Biodiesel Alliance

In the United States, the Sustainable Biodiesel Alliance (SBA) was created to encourage sustainable practices throughout the biodiesel life cycle. "Biofuels now are becoming the gold rush," says Heidi Quante, the Sustainable Biodiesel Alliance's Executive Director. "The speed with which the biofuels industry has exploded has caused a lot of concern over how consumers can determine which biodiesel is sustainable and which is not." The SBA chose to focus exclusively on standards for biodiesel because that is where the expertise of its founders lay.

This non-profit organization is working to create best practices for the biodiesel industry and to offer certification to ensure that biodiesel is being sourced, manufactured, distributed, and used in ways that do little harm to the environment and people. SBA also plans to educate the industry regarding best practices. There was a recognized need for sustainable biodiesel standards because biodiesel is used in existing diesel vehicles. Several sectors of our society will be affected by a shift to biodiesel and other renewable fuels, including school buses, urban buses, and the US commercial trucking fleet. It is important to ensure that our biodiesel is as sustainable as possible.

Working through a multi-stakeholder process, the SBA has drafted sustainable biodiesel principles that will be used in the sustainable biodiesel certification process. For example, the SBA believes that biodiesel feedstocks should be grown and the biodiesel produced and consumed locally. It does not favor importing fuel from other countries – or communities. Ultimately, however, the steering committee will determine the criteria.

As with LEED, certification is voluntary. The government is not involved as it has been in the development of the organic standard. Biodiesel certification simply provides a way for companies to let consumers know that their products are greener throughout the biodiesel life cycle.

The SBA's web site contains a link to the Alternative Fuels Data Center's web site for those wishing to find the nearest biodiesel or other alternative fuels that are available.

Europe: European Biodiesel Board

In 2003, Europe adopted the Biofuels Directive, which called for biofuels to replace up to 2% of Europe's transport fuels by 2005 and up to 5.75% by 2010. Although the EU did not reach its 2005 goal, its 2010 goal still stands; in addition, the EU has established an aggressive goal for biofuels use by 2020.

Biofuels growers and producers have ramped up their operations to meet these goals. However, Europe has realized that it probably cannot provide enough feedstock domestically to meet the growing demand for biodiesel and other biofuels. How will Europe find the feedstocks to meet its biofuels goals? It will buy them from international markets. This is a cause for deep concern due to the non-sustainable production of feedstocks.

The European Biodiesel Board (EBB), Europe's largest association of biodiesel producers, has called for a system to certify greener biodiesel. Established in Brussels, Belgium, in 1997, the EBB promotes biodiesel use and acts as a representative body for its members.

The point of such a certification system would be to "screen out" feedstocks that were produced or are suspected to have been through deforestation or other non-sustainable means (Navarro 2007b). Therefore, such a certification system, or ecolabel, could be used to identify greener imported and domestic feedstock.

Establishing Sustainable Biodiesel Guidelines

Not all organizations working toward sustainable biodiesel are creating ecolabels. Some are creating voluntary guidelines or principles that should be followed in the production of biodiesel. These include the Roundtable on Sustainable Biofuels, high-level Declaration on Sustainable Biofuels Development in Africa, and Roundtable on Sustainable Palm Oil. Each is discussed below.

Worldwide: Roundtable on Sustainable Biofuels

The Roundtable on Sustainable Biofuels has been established to develop a global set of biofuels standards. Created by the Swiss Energy Center at the École Polytechnique Féférale de Lausanne (EPFL), the Roundtable consists of multiple stakeholders, including representatives from Toyota, British Petroleum, the Dutch and Swiss governments, the UN Foundation, the World Economic Forum, Friends of the Earth Brazil, and others. The Roundtable on Sustainable Biofuels is working to create standards that are simple, generic, adaptable, and efficient.

The group's Steering Board has drafted principles for the sustainable production of biofuels. These principles are shown below.

- "Biofuel production should not directly or indirectly endanger wildlife species or areas of high conservation value.
- Biofuel production should not directly or indirectly degrade or damage soils.
- Biofuel production should not directly or indirectly contaminate or deplete water resources.
- Biofuel production should not directly or indirectly lead to air pollution.
- The use of biotechnologies for biofuels production should improve their social and/or environmental performance, and always be consistent with national or international biosafety protocols" (BioFuels Digest 2007).

Four working groups will focus on greenhouse gas lifecycle efficiency, environmental concerns, social concerns, and implementation. Anyone interested in the standards can provide feedback during their development.

The Roundtable will use the "Alliance Code of Good Practice for Standard Setting," developed by the International Social and Environmental Accreditation and Labeling, to establish its sustainable biofuels standards. A draft of these standards is expected to be developed by the middle of 2008.

Africa: Declaration on Sustainable Biofuels Development in Africa

In July and August 2007, the first high-level African biofuels seminar was held in Addis Ababa, Ethiopia. Those involved included the Africa Energy Commission, New Partnership for Africa's Development, and the United Nations Industrial Development Organization. At a Ministerial Roundtable, African ministers and various representatives adopted the Addis Ababa Declaration on Sustainable Biofuels Development in Africa.

Among other things, the declaration called for Africa's participation in the global sustainability dialogue. It established the need to create biofuel principles; reduce risks for small-scale biodiesel producers; and create regulatory and policy frameworks that will support the development and existence of biofuels. The African ministers and other high-level representatives present committed to the declaration.

This type of activity is important because Africa's many non-oil-producing countries no longer can afford the rising costs of fossil fuels. Even in oil-producing countries, poorer residents cannot afford electricity. Guinea's Minister of Agriculture, Livestock, Environment, Water, and Forests suggested that the main goal for biofuels should be to support the rural poor and to improve their lives (IISD 2007). He suggested that the domestic use of biofuels should come first and that export should occur only after Africa's needs have been met.

Worldwide: Roundtable on Sustainable Palm Oil

The Roundtable on Sustainable Palm Oil is a multi-stakeholder organization that works to ensure that the palm oil supply chain is sustainable. Its first formal meeting was held in 2003 in Malaysia where 200 people attended representing sixteen countries (RSPO 2007). The Roundtable has undertaken several activities, including the development of definitions, criteria, and best practices for sustainable palm oil. The organization's principles for sustainable palm oil production are shown below.

"Principle 1: Commitment to transparency

Criterion 1.1 Oil palm growers and millers provide adequate information to other stakeholders on environmental, social, and legal issues relevant to RSPO Criteria, in appropriate languages & forms to allow for effective participation in decision making.

Criterion 1.2 Management documents are publicly available, except where this is prevented by commercial confidentiality or where disclosure of information would result in negative environmental or social outcomes.

Principle 2: Compliance with applicable laws and regulations

Criterion 2.1 There is compliance with all applicable local, national, and ratified international laws and regulations.

Criterion 2.2 The right to use the land can be demonstrated, and is not legitimately contested by local communities with demonstrable rights.

Criterion 2.3 Use of the land for oil palm does not diminish the legal rights, or customary rights, of other users, without their free, prior, and informed consent.

Principle 3: Commitment to long-term economic and financial viability

Criterion 3.1 There is an implemented management plan that aims to achieve long-term economic and financial viability.

Principle 4: Use of appropriate best practices by growers and millers

Criterion 4.1 Operating procedures are appropriately documented and consistently implemented and monitored.

Criterion 4.2 Practices maintain soil fertility at, or where possible improve soil fertility to, a level that ensures optimal and sustained yield.

Criterion 4.3 Practices minimize and control erosion and degradation of soils.

Criterion 4.4 Practices maintain the quality and availability of surface and ground water.

Criterion 4.5 Pests, diseases, weeds, and invasive introduced species are effectively managed using appropriate Integrated Pest Management (IPM) techniques.

Criterion 4.6 Agrochemicals are used in a way that does not endanger health or the environment. There is no prophylactic use, and where agrochemicals are used that are categorised as World Health Organisation Type 1A or 1B, or are listed by the Stockholm or Rotterdam Conventions, growers are actively seeking to identify alternatives, and this is documented.

Criterion 4.7 An occupational health and safety plan is documented, effectively communicated, and implemented.

Criterion 4.8 All staff, workers, smallholders, and contractors are appropriately trained.

Principle 5: Environmental responsibility and conservation of natural resources and biodiversity

Criterion 5.1 Aspects of plantation and mill management that have environmental impacts are identified, and plans to mitigate the negative impacts and promote the positive ones are made, implemented, and monitored, to demonstrate continuous improvement.

Criterion 5.2 The status of rare, threatened or endangered species and high conservation value habitats, if any, that exist in the plantation or that could be affected by plantation or mill management, shall be identified and their conservation taken into account in management plans and operations.

Criterion 5.3 Waste is reduced, recycled, re-used, and disposed of in an environmentally and socially responsible manner.

Criterion 5.4 Efficiency of energy use and use of renewable energy is maximised.

Criterion 5.5 Use of fire for waste disposal and for preparing land for replanting is avoided except in specific situations, as identified in the ASEAN guidelines or other regional best practice.

Criterion 5.6 Plans to reduce pollution and emissions, including greenhouse gases, are developed, implemented, and monitored.

Principle 6: Responsible consideration of employees and of individuals and communities affected by growers and mills

Criterion 6.1 Aspects of plantation and mill management that have social impacts are identified in a participatory way, and plans to mitigate the negative impacts and promote the positive ones are made, implemented and monitored, to demonstrate continuous improvement.

Criterion 6.2 There are open and transparent methods for communication and consultation between growers and/or millers, local communities, and other affected or interested parties.

Criterion 6.3 There is a mutually agreed and documented system for dealing with complaints and grievances, which is implemented and accepted by all parties.

Criterion 6.4 Any negotiations concerning compensation for loss of legal or customary rights are dealt with through a documented system that enables indigenous peoples, local communities, and other stakeholders to express their views through their own representative institutions.

Criterion 6.5 Pay and conditions for employees and for employees of contractors always meet at least legal or industry minimum standards and are sufficient to meet basic needs of personnel and to provide some discretionary income.

Criterion 6.6 The employer respects the right of all personnel to form and join trade unions of their choice and to bargain collectively. Where the right to freedom of association and collective bargaining are restricted under law, the employer facilitates parallel means of independent and free association and bargaining for all such personnel.

Criterion 6.7 Child labour is not used. Children are not exposed to hazardous working conditions. Work by children is acceptable on family farms, under adult supervision, and when not interfering with education programmes.

Criterion 6.8 The employer shall not engage in or support discrimination based on race, caste, national origin, religion, disability, gender, sexual orientation, union membership, political affiliation, or age.

Criterion 6.9 A policy to prevent sexual harassment and all other forms of violence against women and to protect their reproductive rights is developed and applied.

Criterion 6.10 Growers and mills deal fairly and transparently with smallholders and other local businesses.

Criterion 6.11 Growers and millers contribute to local sustainable development wherever appropriate.

Principle 7: Responsible development of new plantings

Criterion 7.1 A comprehensive and participatory independent social and environmental impact assessment is undertaken prior to establishing new plantings or operations, or expanding existing ones, and the results incorporated into planning, management, and operations.

Criterion 7.2 Soil surveys and topographic information are used for site planning in the establishment of new plantings, and the results are incorporated into plans and operations.

Criterion 7.3 New plantings since November 2005 (which is the expected date of adoption of these criteria by the RSPO membership), have not replaced primary forest or any area containing one or more High Conservation Values.

Criterion 7.4 Extensive planting on steep terrain, and/or on marginal and fragile soils, is avoided.

Criterion 7.5 No new plantings are established on local peoples' land without their free, prior, and informed consent, dealt with through a documented system that enables indigenous peoples, local communities, and other stakeholders to express their views through their own representative institutions.

Criterion 7.6 Local people are compensated for any agreed land acquisitions and relinquishment of rights, subject to their free, prior, and informed consent and negotiated agreements.

Criterion 7.7 Use of fire in the preparation of new plantings is avoided other than in specific situations, as identified in the ASEAN guidelines or other regional best practice.

Principle 8: Commitment to continuous improvement in key areas of activity

Criterion 8.1 Growers and millers regularly monitor and review their activities and develop and implement action plans that allow demonstrable continuous improvement in key operations." (RSPO 2006)

These principles are based on the assumption that sustainable palm oil production should be economically, socially, environmentally, and legally viable and beneficial. Many organizations have signed onto the principles. For additional information, visit the Roundtable for Sustainable Palm Oil's web site at http://www.rspo.org/.

The Need for True Sustainability

As was stated in the Introduction, sustainability means that a product or a process does not diminish the integrity or resilience of any ecosystem, human community, or human economy during any stage of its life cycle.

Many of the qualities attributed to sustainable biodiesel – local feedstock production (or recycling of waste oil/fat), local biodiesel production, reuse/recycling of sidestreams, and local biodiesel use – are important. They help to ensure that the community that grows, produces, recycles, and uses local biodiesel benefits economically and socially by keeping every stage of the biodiesel life cycle close to home.

But what about the environment? Sustainability is not just about supporting communities – it also is about the world's ecosystems upon which all communities rely for every element of their existence. Keeping it local "from feedstock to fuel" is far superior – and more sustainable – to the industrial model of agriculture and production that pervades much of the world's farmland and industry. Yet, while small may be beautiful because it leaves a smaller ecological footprint, it is not inherently sustainable.

Let's assume that biodiesel is made using a first-generation (virgin) feedstock (as opposed to a second-generation feedstock, such as waste oil/grease). Most farming is inherently non-sustainable. Unless they are created to mimic natural ecosystems, farms are never as ecologically beneficial as the habitat that they replaced. As a result, natural capital is depleted, and local human economies are limited as are the evolutionary opportunities of native species and communities.

Sustainable farming should be based on the way that nature "farms" – organically, using a broad diversity of mixed crops (preferably native), with no tilling (this breaks up the ecological communities that live in the soil and that build soil), harvested by hand or using methods that do not tear up the soils, and appropriately scaled. Although genetically modified organisms (GMOs) are a reality in modern farming, there are many ecologically based arguments against them. For now, they have a place in biodiesel production simply because there is so much GMO feedstock available, but this issue should be revisited from a scientific standpoint rather than a business or political one.

Then, the feedstock needs to be transported to the biodiesel production facility, which requires energy. The finished biodiesel also needs to be moved from production facility to distribution sites or to the end users. Are the production facilities entirely green, giving back as much as they take ecologically? Is fossil energy used to accomplish this work, or is the energy made from renewable *and* environmentally friendly sources?

Biodiesel production requires energy and produces waste products. What energy powers the production of the biodiesel? What happens to the waste products? When it is used, biodiesel might emit fewer pollutants than does petroleum diesel, but still it emits pollutants.

Lyle Estill of Piedmont Biofuels suggests, "If we are to achieve sustainability, we will need as many "closed loop" relationships as possible." This means keeping biodiesel production appropriately sized. It also means using only those resources that our local communities and economies can provide sustainably while keeping production and use local to support the local economy.

However, it also means more than that. It means developing and using truly sustainable – not just small-scale – farming practices. It means ensuring that biodiesel production facilities and production processes are sustainable – not just energy-efficient but causing "zero net impact," or, better yet, being ecologically regenerative. It means storing, packaging, and transporting biodiesel in a way that does no harm. Is it possible to use biodiesel sustainably since it is a polluting substance – albeit better than most of our current fuel options?

If we are not truly committed to sustainability but simply want to reduce our environmental harm, then we will need to find a more suitable term for our biodiesel principles and criteria than "sustainable."

Chapter 8. Conclusion

This chapter provides a brief summary of the book's key points. It touches on some of the potential benefits of biodiesel. Then, it discusses some of the challenges that biodiesel producers face. Finally, it summarizes the importance of bringing sustainability to the center of the biodiesel dialogue.

Biodiesel is attractive for many reasons. Its emissions are environmentally preferable to those of some other fuels, including petroleum diesel. It also can be sourced, produced, and used domestically and is an easy replacement for petroleum diesel. Biodiesel is engine friendly. When virgin feedstocks are used, it can be carbon neutral; when waste cooking oil is used, it provides a way to recycle waste oil. When externalities are factored in, biodiesel can be relatively inexpensive to produce. It also has many uses and can stimulate the domestic economy. Biodiesel production is not energy intensive.

However, not all biodiesel is created equal. Once viewed as an environmentally preferable renewable fuel, some now view biodiesel and other biofuels with suspicion as nations around the globe are rushing to meet the world's energy demands with their oil feedstocks. Unfortunately, some of those feedstocks are grown in ways that are ecologically, economically, and culturally harmful. The social and environmental dangers inherent in non-sustainable biodiesel oilseed crop production threaten ecosystems overseas and in the United States as ever-growing areas are cleared for biofuels production.

As biodiesel increases in popularity in the US and overseas, it is of the utmost importance that it remain an environmentally preferable fuel throughout its entire life cycle. The replacement of habitat with biodiesel monocrops results in the loss of carbon sequestering biomass and the ecological benefits of more complex ecosystems. The forced removal of people from their land and falsification of land deeds is a severe human rights violation that flies in the face of some of the qualities that make the idea of biodiesel so attractive, such as the strengthening of local economies and communities.

The growth of the biodiesel industry has been remarkable, and many of these issues have emerged quickly. However, these are issues that will need to be resolved before biodiesel can be considered a greener fuel. Although best practices and regulations have been developed to address some of the immediate environmental and human health associated with biodiesel, a more holistic, larger-scale set of biodiesel best practices, regulations, and incentives are desperately needed to address some of these issues.

For example, the biodiesel industry is working to address issues of safety, storage, and waste disposal. Because methanol is flammable and can harm human health, safety precautions, storage protocols, and disposal regulations must be followed. Biodiesel and glycerol also must be stored properly. Regulations vary by state regarding the disposal of glycerin and wastewater, and some states have developed specific regulations that outline the disposal options for glycerin/glycerol and wastewater. These are all important at the individual producer's level, but they do not address safety and waste disposal on local, regional, national, or international scales.

Similarly, the American Society of Testing and Materials created the ASTM D 6751 standard to ensure that biodiesel sold and used is of good quality. Furthermore, the IRS has fuel compliance officers that inspect fuel quality nationwide, and the EPA requires that all commercial biodiesel be registered and meet the Clean Air Act standards. Some state agencies also regulate biodiesel although regulations vary widely. The development of voluntary quality standards, such as BQ-9000, also is important. Yet, for one concerned with sustainability, biodiesel grown on farmland that replaced a tropical forest or on land taken after its inhabitants were forcibly removed would be considered poor quality – even if it did pass the ASTM specification. Thus, the existing quality standards are fine for what they are and do what they are intended to do; however, they do not do enough to encompass sustainability or address it in any real way.

Since biodiesel and other alternative fuels are fairly new to most people, there are few biodiesel-specific regulations in place. Those that exist can be difficult to find and vary by state. Increased attention and interest in biodiesel calls for fast action by states' regulatory bodies to develop regulations that make sense, do not hinder small-scale energy projects, and are easy to follow while ensuring that current and future biodiesel producers follow the proper guidelines so that good intentions do not end in disaster. However, most existing environmental regulations focus not on issues of global sustainability but on very localized guidelines that vary from place to place and might include several local authorities, from fire marshals to environmental health departments.

In response to the development of the biodiesel industry, best practices have been established and regulations and incentives have been created to protect human health and guard against environmental harm. However, as we have explained, they do not go far enough to achieve sustainability. Clearly, we need to bring sustainability to the center of the biodiesel dialogue before biodiesel production has done more harm than good.

If we really are committed to sustainable biodiesel, it will need to be made using a variety of sustainably grown and harvested, rapidly renewable feedstocks that are locally grown and appropriate to their region. It will need to be produced on a scale that is appropriate to the amount of local source material and the size of the local market. As a result, it will help to support local farms and economies over the long term as long as the feedstocks are being produced sustainably. Thus, biodiesel will be a part of community building or rebuilding. Biodiesel that is made this way can be certified so that consumers know that what it is they are purchasing is environmentally preferable.

Sustainability also will require us to change by using less and conserving more energy. Although conservation and lifestyle change do not appear to be popular topics, there is no way that existing biofuels can meet our current energy demands. We will have to step down our energy use as we transition to renewable, environmentally friendly fuels.

Biodiesel can play an important role in America's energy portfolio. However, it must be appropriately scaled and thoughtfully manufactured, transported, and used – in concert with a reduction of our energy demands – to ensure that it does not do more ecological harm than good. Our hope is that, as the use of biodiesel increases, a cohesive body of best practices, regulations, and incentives will be put into place that will protect human and environmental health and safety while providing an alternative to fossil fuels. If sustainable biodiesel is really about appropriate scaling and local sourcing, production, and use, then these best practices, regulations, and incentives will need to support – rather than stymie – small and medium-scale, local production.

Glossary of Terms & Acronyms

AB – agri-biodiesel, which is made of virgin feedstock

AFDC – Alternative Fuels Data Center

ANWR – Alaska National Wildlife Refuge

Aromatic compound – a chemical compound that has an odor or flavor or that contains aromatic rings, such as benzene.

ASA – American Soybean Association

ASTM – American Society for Testing and Materials

ASTM D 6751 – the standard used for pure biodiesel (B100)

B (as in *B100, B20*) – the number behind "B" is the percentage of biodiesel in the fuel blend

Biodiesel – a type of fuel made that typically is made using vegetable oil/animal fats that have been chemically altered using methanol and a catalyst, such as lye. Researchers also are working to make biodiesel using algae as the oil source. The National Biodiesel Board's technical definition of biodiesel is as follows: "a fuel comprised of mono-alkyl esters of long chain fatty acids derived from vegetable oils or animal fats, designated B100, and meeting the requirements of ASTM D 6751" (NBB 2007i).

Biodiesel blend – the National Biodiesel Board's definition is "a blend of biodiesel fuel meeting ASTM D 6751 with petroleum diesel fuel, designated BXX, where XX represents the volume percentage of biodiesel fuel in the blend" (NBB 2007i).

Bioenergy – energy that is produced from biological ingredients

Biofuel – a fuel, such as biodiesel, vegetable oil, or ethanol, that is made from biological ingredients

Bioheat – biodiesel that is used for heating a facility (rather than for powering a vehicle)

BOD – biochemical oxygen demand

BQ-9000 – a biodiesel quality standard

Btu – British thermal units

Carbon monoxide – a gas made of bonded carbon and oxygen atoms in a 1:1 ratio

Carbon dioxide – one of the gases that affects the global climate

Cloud point – the temperature at which paraffin crystals begin to form in oils

CO2 – carbon dioxide

COD – chemical oxygen demand

DEQ – Department of Environmental Quality

Diesel – fossil-based (petroleum) diesel

DOE – Department of Energy

DOT – Department of Transportation

EPA – Environmental Protection Agency

Ethanol (also ethyl alcohol or grain alcohol) – a chemical compound that can be used as a fuel. It is made using various biological feedstocks, such as corn or sugar.

Feedstock – primary oil ingredient used to make biodiesel

FIE – fuel injection equipment

Flash point – the temperature at which the biodiesel burns

Fossil fuel – oil, coal, or natural gas

FFA – free fatty acids

FOG – fats, oil, and grease

GDP – gross domestic product

GHG – greenhouse gas

Glycerin – a byproduct of biodiesel production

Glycerol – a byproduct of biodiesel production, which contains glycerin and methanol

Hydrocarbons – an organic compound, made of hydrogen and carbon, that causes air pollution

KOH – potassium hydroxide

Methane – a gas that affects the global climate

Methanol (also methyl alcohol or wood alcohol) – a chemical compound that forms the simplest alcohol and is used in making biodiesel

Methoxide – a combination of methanol plus a catalyst, such as lye

MSDS – materials safety data sheet

NB – biodiesel, which is made using non-virgin feedstock

NBB – National Biodiesel Board

NOx – nitrogen oxides

NPAH – nitro polycyclic aromatic hydrocarbons

NREL – National Renewable Energy Laboratory

Original engine manufacturers (OEMs) – generally, this is the company that manufactures the vehicle

Ozone – a molecule that consists of three oxygen atoms and acts as a pollutant in the biosphere

PAH – polycyclic aromatic hydrocarbon, a chemical compound that is a pollutant

Particulates/particulate matter – very small particles of a liquid or solid that are suspended in a gas

Peaker power plant – a power plant that runs only when electricity demand is high

Petroleum diesel – fossil fuel-based diesel fuel

Potassium hydroxide (KOH) – one of the catalysts used for making biodiesel

POTW – publicly owned treatment works

Pour point – the lowest temperature at which oil will flow

RCRA – Resource Conservation and Recovery Act

RFS – renewable fuel standard

SSP – small-scale producer

Small-scale producer – for our purposes, biodiesel producers who range from home brewers up to those who make up to 250,000 gallons per year

Smog – a type of air pollution originating from the combined terms "smoke" and "fog"

SO_2 – sulfur dioxide

SOx – sulfur oxides

Speciated hydrocarbon – a type of pollution

Sulfates – a chemical compound made of sulfur and oxygen atoms

Sulfur oxides – includes sulfur dioxide, sulfur trioxide, which are pollutants

SBA – Sustainable Biodiesel Alliance

SVO – straight vegetable oil

Tier I and Tier II – health effects testing requirements developed under the US Environmental Protection Agency's Clean Air Act amendments

transesterification – a chemical reaction created using fat, alcohol, and a catalyst

ULSD – ultra-low sulfur diesel

USDA – US Department of Agriculture

Appendix A. Biodiesel Production Around the World

This section discusses the growing popularity of biodiesel around the world. It provides a high-level look at some of the countries that are producing biodiesel, focusing specifically on the incentives, production, markets, and main feedstocks that those countries are using. Then, it examines some of the ecological problems that are resulting from the growth of oilseed plantations.

The Global Growth of Biodiesel Production

Biodiesel is gaining in popularity worldwide. While many countries are producing and using or exporting their biodiesel, others are in the planning stages for biodiesel production and distribution. This chapter explores some nations' efforts at producing and using or exporting biodiesel. Because much of the information that is publicly available relates primarily to large-scale production, this chapter will serve as a general discussion of biodiesel around the world rather than focusing exclusively on small-scale production. It provides an overview of the incentives, biodiesel production and markets, and oilseed crops in several nations around the world.

Some of the countries that are involved in biodiesel production include:

- Argentina
- Australia
- Belgium
- Brazil

- Canada
- Colombia
- Czechoslovakia
- Finland
- Germany
- India
- Indonesia
- Japan
- Malaysia
- Malta
- Papua New Guinea
- Singapore
- Thailand
- United Kingdom
- Zambia

This list is by no means comprehensive as several other individuals, companies, communities, and governments are producing or have plans to produce biodiesel. For example, in Montenegro, the Victoria Group plans to produce 100,000 tons of biodiesel annually (Renewable Development Initiative 2006). In Costa Rica, some biodiesel reactors are made in Costa Rica and sold throughout the Caribbean and Central American areas. Because Costa Rica produces palm oil, there is an interest in producing biodiesel.

In Romania, two biodiesel projects are under way as are projects in Latvia, Bulgaria, and Ukraine. Sweden's Statoil is selling B5 biodiesel at its fueling stations. In Hungary, biodiesel plants are planned for the towns of Babolna and Hodmezovasarhely. A Slovenian oil refinery has partnered with an Austrian company to produce the largest biodiesel refinery in Slovenia that will produce enough biodiesel to meet 88% of the country's demands (STA 2007). A Norwegian oil company, Hydro-Texaco, sells biodiesel in

Oslo, Hadeland, and Lillehammer (Wikipedia 2007a) while Milvenn, another company, sells biodiesel in Bergen.

In New Zealand, Aquaflow Bionomic Corporation, a biodiesel developer, is working to raise money to develop biodiesel from wild algae. Poland's government passed biodiesel regulations allowing the sale of methyl ester biodiesel. Many believe that this will boost the production from 70,000 in early 2006 to 500,000 tons by 2010 (Krukowska 2006).

Largely because of their ability to grow palm oil, many Asian nations are manufacturing biodiesel for their use or for export. The figure below indicates some of the countries that were using biodiesel in 2007.

The market for biodiesel is strong in Europe, where the Parliament offered a 90 percent biodiesel tax exemption in 1994 (Canadian Renewable Fuels Association 2006). This, combined with alternative fuel legislation and subsidies for growing oilseed crops, has resulted in a strong biodiesel market.

The table below indicates some countries that are involved in biodiesel production and some of the oilseed crops that they are using.

Table 3. Biodiesel Production & Oilseed Crops in Several Countries

Country	Oilseed Crop(s)
Argentina	Soybeans
Australia	Canola, mustard
Belgium	Rapeseed (canola)
Brazil	Castor beans, soybeans, sunflower seeds
Canada	Canola, tallow
Colombia	Palm oil
Czechoslovakia	Rapeseed

Finland	Rapeseed
Germany	Rapeseed
India	Palm, soy, *Jatropha curcas* (Ratanjyot), Pongamia pinnata (Karanj), Calophyllum inophyllum (Nagchampa), Hevcca brasiliensis (Rubber)
Indonesia	Palm, jatropha??
Japan	Recycled oil
Malaysia	Palm oil
Malta	Waste vegetable oil/animal fats
Papua New Guinea	Copra (coconut) oil
Singapore	Palm
Thailand	Palm oil, coconut oil, waste vegetable oil, jatropha
United Kingdom	Rapeseed (canola)
Zambia	Jatropha

Indonesia	Japan	Malaysia	Philippines	South Korea
World's second largest palm oil producer and exporter	B100 standard voluntary	World's largest palm oil producer and exporter	World's largest coconut oil producer	B5, B20, and B100 standards set
B100 standard set	B5 standard effective March 2007	National Biofuel Policy looking at B5 - and at export markets	B1 blends introduced in government fleets in March 2004	Refineries supply 0.5% of biodiesel volume nationwide 2004-2006
Goal to replace 10% of conventional fuel use with biofuels by 2010		Plans to build many palm biodiesel plants	Biofuels Act passed in January 2007 requiring B1 within three months and B2 within two years of act's start date	
B5 and B10 sold in selected stations in May 2006; selling 2.5% biodiesel blends by volume in 2007		B5 trials began in selected government vehicles in March 2006		

Biodiesel usage in Asia (2007). *Adapted from IFQC 2007.*

The remainder of this chapter explores the incentives, production and markets, and feedstock used by several nations around the world.

Country-by-Country Incentives, Production, Markets, and Feedstocks

Argentina

Incentives: In 2006, Argentina's government passed a law requiring that 5% of all fuels must comprise biofuels by 2010. Furthermore, the law provides tax incentives to producers of biodiesel and other biofuels while guaranteeing them market share for fifteen years (Valente 2006b). The government will retain the authority to give tax benefits to companies it chooses, giving priority to farmers, small and medium companies, and regional economies (Valente 2006b).

These factors, as well as the fact that production costs are low and demand is high, make biodiesel a promising substitute for petroleum diesel in Argentina. This has opened the door for companies to enter the market for biodiesel production. Currently, the country's farmers use about three-quarters of the biodiesel that is produced.

Production & Markets: For example, the Spanish-Argentinian petroleum company Repsol YPF is spending $30 million to create the Argentinean Bio Fuel Investigation Center, which is projected to produce 120,000 cubic meters of biodiesel annually (Renewable Energy Access 2006). The Center is slated to open in 2007, and its product is expected to be combined with petroleum diesel to result in B5.

Another company, Imperium Renewables, also will produce biodiesel in Argentina. This Seattle startup company will spend approximately $50 million to build its production facility and anticipates making about 100,000 million gallons of biodiesel annually (Kaihla 2006).

Smaller scale producers have become involved with biodiesel as well. For instance, industrial engineer Edmundo Defferrari created a plant for $150,000 that makes 130,000 gallons of biodiesel annually (Kaihla 2006). He sells biodiesel to local farmers, charging about two-thirds of what they pay for petroleum diesel. Olifox, another small company, has already exported biodiesel to Germany but has not been able to meet demand. Biofuels SA, founded by engineer/economist Ricardo Carlstein, supports the market for small-scale, decentralized processing with a new device that will produce biodiesel easily.

Feedstock: Soybeans are Argentina's top export product and provide the feedstock for the country's biodiesel production. From 1996-2004, soybeans were farmed on nearly half of the country's arable land (Valente 2006a).

Australia

Incentives: Australia's government established legislation to support the production and use of biodiesel in 2003. By 2006, it had created the Fuel Tax Bill in which subsidies were provided to biodiesel producers; as a result, no excise taxes will be levied on biodiesel sales until 2011. The federal government is giving partial funding to a nearly $1 million initiative for research into the production of biodiesel from algae (Irani 2006).

Production & Markets: Biodiesel is being produced by both small-scale producers, like BE Bioenergy, and large-scale producers, such as the Australian Biodiesel Group Limited, a publicly traded company.

Every year, Australia uses more than 3.69 billion gallons of petroleum diesel, and the country's nascent biodiesel industry is growing (ABG Biodiesel 2007). In 2006, Gull service stations began offering B20 biodiesel to consumers. The company reeFUEL sells about 13,000 gallons/week of B100 in Townsville, North Queensland. In South Australia, the company Australian Farmers Fuel began selling its B100 to the public in 2001 and sells B20 ("Premium Diesel") at many service stations (Wikipedia. 2007a). Biodiesel also is available in Marrickville (outside Sydney). The city of Adelaide already runs all of its metropolitan buses and trains on B5 and plans to switch to B20.

Feedstock: Canola and mustard (*Brassica juncea*) are used for production although imported palm oil might be used for larger plants. The South Australia Research and Development Institute (SARDI) has started a program for developing canola and mustard seeds for biodiesel production in South Australia. SARDI also is conducting research into making biodiesel from algae.

Belgium

Incentives: Prior to 2006, the government did not allow the manufacture of biodiesel. Now, however, the government of Belgium has passed legislation that will provide subsidies to biodiesel producers. Furthermore, the excise tax levied on the sale of biodiesel has been suspended until at least 2.45% of biodiesel is blended with petroleum diesel.

Production & Markets: Belgium's first biodiesel producer was the company Oleon, which sells its biodiesel through the supplier Total. In 2005, about 20,000 metric tons of biodiesel were produced in Belgium and the Netherlands combined. Refineries are located in Ertvelde and at Feluy. Total expects that biodiesel sales in Belgium will reach the minimum 5.75% of the fuel market by 2010 as set out by the EU. Three new companies began biodiesel production in 2007 and will give Belgium a total annual capacity of 350,000 tons. A new partnership between Cargill and two Belgian companies will result in a rapeseed biodiesel plant that is expected to produce 200,000 tons of fuel annually. It is anticipated that both soybean and palm oils also will be used in the country's production of biodiesel as demand for the fuel grows.

Feedstock: Most of Belgium's biodiesel will be made using grain and oilseeds imported from other countries in the European Union. Currently, rapeseed is the preferred oilseed that has been used in Belgium's biodiesel production.

Brazil

Incentives: The government of Brazil has long encouraged the use of biofuels, so its interest in biodiesel is not surprise. In 2002, the government's Ministry of Science and Technology created the Brazilian Programme for the Technological Development of Biodiesel – otherwise known as PROBIODIESEL. In July 2003, the government created a new anti-ministerial working group, which was responsible for assessing Brazil's biodiesel production and use potential. That group determined that biodiesel was, in fact, viable. Law 11097 was passed in January 2005 to define federal guidelines for biodiesel production and use. In May 2005, the government passed Law 11116, which provided tax incentives *that vary by feedstock and region* and is particularly expected to benefit those living in the poorest regions of Brazil (de Sousa, n.d.). By the end of 2006, the government stated that it would require that all petroleum diesel contain a minimum of B5 by 2010.

Production & Markets: In 2005, a commercial biodiesel refinery went into production that can make up to 12,000 cubic meters (3.2 million US gallons) each year (Wikipedia. 2007a). That year, three refineries had been opened, and an additional seven that were slated to open. Together, these three refineries could produce over 12 million litres of biodiesel annually. This biodiesel will be blended with petroleum diesel to create a minimum of B2 until 2010 after which it will be used to make B5. Brazil's national petroleum company, Petrobas, manufactures biodiesel using a petroleum refinery.

Feedstock: The biodiesel feedstock used includes castor beans, soybeans, and sunflower seeds. Of these, castor beans are considered the best because they are less costly and easier to plant (Wikipedia. 2007a).

Canada

Incentives: Canada currently offers no tax incentives for the production or use of biodiesel. However, the Provincial Government of Nova Scotia heats some public buildings using biodiesel and uses it in some of its public transport vehicles.

Production & Markets: Small-scale biodiesel production in Canada has been carried out by individuals and farms, some of whom buy BioFuel Canada Ltd.'s equipment. Biodiesel cooperatives, such as British Columbia's Island Biodiesel Co-op, Vancouver Biodiesel Co-op, and WISE Energy, also produce fuel on a small scale.

Companies such as Biox and Rothsay, which began production in 2006, also have produced biodiesel on a fairly small scale. Biox Corporation's biodiesel processing plant is in Ontario, and Rothsay's production in Quebec yields 35,000 cubic meters of biodiesel annually (Wikipedia. 2007a). Bifrost Bio-Diesel and others built biodiesel plants in Manitoba beginning in 2005.

On a larger scale, Cascadia Biofuels, a joint venture of Autogas Propane Ltd. and United Petroleum Products Inc., distributes a biodiesel mix through some of its retail stations in British Columbia.

Wilson Fuels, which uses waste fish oil to produce biodiesel, has opened a biodiesel station in New Brunswick. Retail fueling stations also exist in Toronto and Unionville. The bus fleet owned by Halifax Regional Municipality has been converted to run on biodiesel. Future demand is projected to be 3,000 cubic meters of B20-B100 for buildings and 7,500 cubic meters of B20-B50 for transport fuel

(Wikipedia. 2007a). In Nova Scotia, biodiesel is used to heat public buildings and, to a lesser extent, for public transport.

Feedstock: Canadian biodiesel relies primarily on two feedstock sources – canola oil and tallow. Historically, Canada's canola crops have been used for oil, with exports mostly to Japan, and for meal, for domestic use or export primarily to the US. The high cost of canola production is an obstacle to widespread, domestic biodiesel production in Canada.

Colombia

Incentives: In 2004, Law 939 was passed stating that Colombian petroleum diesel must include five percent biodiesel. A similar law was passed in 2001 in support of ethanol production and use.

Production & Markets: Despite that rapid growth of Colombia's ethanol industry, the use of petroleum diesel oil for vehicles is rising at a rate higher than that of gasoline. The demand for petroleum diesel has surpassed supply, so Colombia imports about five percent of its petroleum diesel oil.

African palm trees are already under cultivation with 600 thousand tons of oil being produced annually. However, the conversion of forest land to palm oil plantations is rising quickly. In the mid-1960s, African palm grew on about 18,000 hectares of land. In 2003, African palm trees grew on more than 188,000 hectares. A total of about 300,000 hectares have been planted, and the government hopes to have one million hectares under palm cultivation soon (Avendaño 2006). Several palm oil plants have been built around Colombia. Currently, Colombia is the world's fourth largest producer of palm oil and its largest producer in the Americas. Only Indonesia, Malaysia, and Nigeria surpass Colombia in palm oil production, about one-third of which is exported.

Feedstock:
Colombia's primary oilseed feedstock is the African palm.

Czech Republic

Incentives: In 2004, the Czech Republic's government began subsidizing biodiesel production by reducing the value-added tax on biodiesel. Furthermore, the warranties on all Skoda diesel vehicles manufactured since 1996 approve the use of biodiesel.

Production & Markets: The Czech Republic began producing biodiesel more than a decade ago, resulting in 60,000 cubic meters per annum by the early 1990s (Wikipedia. 2007a). Much of this is based on large-scale production, including one plant in Olomouc that makes nearly 40,000 cubic meters annually (Wikipedia. 2007a).

Feedstock: Rapeseed is an important crop in the Czech Republic and is a natural oilseed for biodiesel production.

Finland

Incentives: There are few incentives in place for the production of biodiesel in Finland. Taxation policies and the costs of production have kept biodiesel production and use from being economically competitive.

Production & Markets: In 2007, Neste Oil began making alkyl biodiesel through what it refers to as the NExBTL process. This alkyl biodiesel is an oxygenless paraffin with a minimal carbon dioxide output throughout its life cycle. The company's annual production capacity is 170,000 tons (Wikipedia. 2007a). This will make Finland the fourth largest European biodiesel producer after Germany, France, and Italy (Green Car Congress 2005). The French company Total plans to build some biodiesel refineries in Finland in 2008.

Feedstock: Finland has about 30,000 hectares of land set aside for non-food rapeseed production, which would result in about 10,000 tons of rapeseed oil each year.

Germany

Incentives: Biodiesel was tax free until August 1, 2006. The renewed tax on biodiesel has made it less competitive with petroleum diesel at the pump, and many large transport firms have returned to petroleum diesel use. However, the European Union also passed legislation, the Biofuels Directive, requiring a maximum of 5% biodiesel to be added to petroleum diesel starting in 2004. It is unclear what affect that will have on Germany's biodiesel market.

Production & Markets: Biodiesel production and use have risen significantly in Germany, making Germany the world's largest biodiesel producer. Between 2002-2004, biodiesel sales doubled to 376.6 million litres, or over 99 million US gallons, which is enough to fuel more than 300,000 cars (Wikipedia. 2007a). In 2004, nearly half (45%) of Germany's biodiesel sales were to trucking companies and other large consumers (Wikipedia. 2007a). In 2006, the country's annual production capacity was projected to be more than 528 million gallons (Wikipedia. 2007a).

Due to their relative efficiency and the rising cost of oil, nearly half of the new cars in Germany are made with diesel engines. This creates a fairly steady market for biodiesel. Currently, about one in ten gas stations sells biodiesel to the public.

Feedstock: Most of Germany's biodiesel is produced using rapeseed.

India

Incentives: In June 2006, India's president A.P.J. Abdul Kalam established energy independence as a national priority and called for the production of 60 million tons (18 billion gallons) of biodiesel annually (Green Car Congress 2006). At the time, India used about 120 millions tons of oil each year – most of it for transportation – and this increased production would meet about half of the country's fuel needs (Green Car Congress 2006). Meanwhile, the government is considering not imposing an excise tax on biodiesel sales in order to encourage use of the fuel.

In 2007, the Indian research organization The Energy and Resources Institute (TERI) began a ten-year study of jatropha as a biodiesel feedstock. TERI has worked to educate farmers in Andra Pradesh about jatropha cultivation. In order to encourage those farmers to grow jatropha, the organization also has worked with local banks to develop loan guarantees for the purchase of seed and to protect farmers against losses. In 2007, about 5,000 farmers had signed up to grow jatropha on about 1,000 hectares of land, and TERI hoped that 8,000 hectares would be under jatropha cultivation by March 2008 (Fitzgerald 2006). Ultimately, TERI hopes to see the manufacture of over 23 million gallons of biodiesel every year.

Meanwhile, the government of Andhra Pradesh created a biodiesel policy that would encourage the state's farmers and investors to grow biodiesel on 1.5 million acres of land. Andra Pradesh expects to purchase the jatropha seeds from contract farmers, thereby helping to reduce farmers' financial risk.

Aware of TERI's efforts, the national government is considering a national program of jatropha cultivation. Other biodiesel incentives are being developed as well. For example, the Petroleum Conservation Research Association has created the Bio-Diesel Credit Bank, which will coordinate carbon credit activities. Tata Motors and Indian Oil also are considering using biodiesel made from jatropha oil.

Production & Markets: Biodiesel is being made on a small-scale and used in motorized rickshaws. Furthermore, the Indian Oil Corporation has purchased biodiesel in order to test it for potential use in railways and vehicles.

In the state of Karnataka, biodiesel is used in state-run buses (Waterloo School for Community Development. n.d.). In the state of Orissa, volunteers from the village of Kinchlingi, which comprises sixteen homes, has made biodiesel since November 2004. The volunteers, who are organized by the local village council, make one five-litre batch of biodiesel weekly. This biodiesel powers a water pump that supplies about 70 litres of water each day to the village's seventy-three residents (Waterloo School for Community Development. n.d.). Such biodiesel production programs for the purpose of pumping water are being implemented in other villages as well.

A company in Gujarat, Aatmiya Biofuels Pvt. Ltd. produces about 1,000 litres of biodiesel daily. Kochi Refineries Ltd. has partnered with a US firm to produce 100 litres of biodiesel daily using the oil from rubber plant seeds. Meanwhile, Bhoruka Power Corporation Ltd. is researching the use of Pongamia or Neem seeds in India's Karnataka state. Shirke Biohealthcare Pvt. Ltd. is establishing a biodiesel refinery in Hinjewadi to produce 5,000 litres of biodiesel daily using jatropha. These are only a few of many biodiesel projects around India.

Feedstock: Several oilseed crops can be used for India's biodiesel production. *Jatropha curcas* (Ratanjyot), Pongamia pinnata (Karanj), Calophyllum inophyllum (Nagchampa), Hevcca brasiliensis (Rubber), palm, and soy are all possibilities. For instance, there are plans to grow jatropha plants on unused land for biodiesel production. Jatropha yields four times the oil that soybeans produce and ten times the amount produced by corn, so it is seen by many as a promising oil crop for biodiesel (Fitzgerald 2006).

In Orissa, oil seeds that grow in local forests are being tested in biodiesel production. These include mahua (Madhuca indica), karanja (Pongammia pinnata), and kusuma (Schleichera Oleosa), which are collected from local forests. Other plants are being grown both in kitchen gardens and agricultural lands. These include niger (Guizotia abyssinica) and castor beans (Ricinus communis).

Bangalore's University of Agriculture Sciences is working to identify helpful strains of oilseed, such as Jatropha curcas and Pongamia pinnata. D1 Oils and Williamson Magor are working to develop large-scale plantations for growing these species in the hilly regions of northeast India and Jharkhand (Wikipedia 2007a).

Indonesia

Incentives: Several laws have been created to encourage the expansion of palm oil plantations. Unfortunately, the rights of indigenous peoples are being abused, and their lands are being taken for use as palm oil plantations. For more information, see *Promised Land: Palm Oil and Land Acquisition in Indonesia: Implications for Local Communities and Indigenous Peoples* (2006).

Production & Markets: Indonesia is the world's largest producer of palm oil with over 6 million hectares of land planted and a total of 18 million hectares of forest cleared for oil palm. Another 20 million hectares have been set aside for oil palm plantations, and there are talks over creating a 1.8-million hectare plantation – the world's largest – in Borneo.

Each year, the Eterindo Group manufactures about 120,000 ton of biodiesel, exporting much of it to the US, Japan, and Germany (Wikipedia 2007a).

Feedstock: Eterindo Group is producing biodiesel using palm oil derivatives.

Japan

Incentives: In order to curb its greenhouse gas emissions and reduce its dependence on foreign oil, Japan announced its goal to run forty percent of its cars on biofuels by March 2009 (AFX News Limited 2006). Because Japan uses far more gasoline than petroleum diesel fuel, these biofuels will focus mainly on a mixture of gasoline and ethanol, which will be produced using sugar grown on an island in the Okinawa chain of islands. Nonetheless, in March 2007, the Japanese government instituted a law requiring up to 5% of biofuel to be blended with petroleum diesel fuel. This law also provides specifications as to biodiesel quality. The government also developed a separate set of voluntary standards for biodiesel.

Production & Markets: Japan began exploring the viability of biodiesel production in 1995. Plans were in the works for a biodiesel plant to be constructed near Tokyo. The plant was expected to produce about .2 million tons of biodiesel annually using recycled oil. Most of Japan's biofuel feedstocks will come from Malaysia or Brazil (Kao 2006).

Feedstock: Recycled oil

Malaysia

Incentives: Malaysia's Deputy Prime Minister has announced, "All efforts will be carried out by the goverment [sic] to promote the development of biodiesel in the country to reach the target of becoming the world's biggest producer of biodiesel" (Clean Air Initiative for Asian Cities 2007). The Deputy Prime Minister cited Europe's desire for biofuels to be a great incentive for biodiesel expansion since Europe was able to meet only one-sixth of its biodiesel demand (Clean Air Initiative for Asian Cities 2007).

Production & Markets: In 2006, Malaysia's prime minister launched a biodiesel program titled Envo Diesel. Envo Diesel is a B5 blend that relies on palm oil as a feedstock. Three palm oil plants are planned that are expected to produce over 500,000 of biodiesel tons in 2007.

Feedstock: Malaysia is the world's second largest producer of palm oil. About eighty-seven percent of the country's recent deforestation is attributed to the establishment of palm oil plantations (World Rainforest Movement 2006).

Malta

Incentives: Malta's government is considering offering an incentive to increase the use of biodiesel in public transportation. Furthermore, the Ministry for Resources and Infrastructure uses biodiesel in its fleet.

Production & Markets: Edible Oil Refining Corporation (EORC), a private corporation that produces 1/3 of Malta's food oils, was the first company in Malta to produce biodiesel (Edible Oil Refining Corporation 2007). Edible Oil's biodiesel refining using WCO began in mid-2002 and increased to 1.8 million liters production in 2006 (Edible Oil, personal communication, 2006).

Edible Oil processes 1,500 tons of fats and oils annually into biodiesel. Of the 1,500 tons of feedstock, 300 tons of virgin oil and 300 tons of waste oil are imported (Edible Oil, personal communication, 2006). The removal of waste oil from the water system is highly desirable given that the cost of clogging due to waste cooking oil (WCO) is estimated at up to $1 million Lm per year (It is not possible to determine, based on existing data, what portion of this is due to motor oil and other contaminants being improperly disposed of (Personal communication with Malta's Minister of the Environment).

Moreover, Malta currently imports waste oil as feedstock for biodiesel production, so the use of domestic waste oil would be advantageous. To capitalize on this resource, a pilot household WCO recycling program was launched in mid-2005 in seven districts. Of the approximate 16,000 households targeted, it was estimated that less than 2,000 had participated as of May 2006 (Edible Oil, personal communication, 2006).

Currently, there are a few fueling stations on Malta that sell biodiesel, and they sell blends ranging from B2 to B30. There is one B100 pump that is located at the Edible Oil's factory where customers may go to fill up their tanks. Biodiesel sells for Lm 0.315 per liter in Malta, which is approximately Lm 0.13 cents cheaper than petroleum diesel. This price of biodiesel is expected to go down in the near future.

Feedstock: Malta's biodiesel is produced using waste vegetable oil and animal fat. Between April 2004 and November 2006, 2,600 tons of waste cooking oil and fat have been collected from Malta's restaurants and processed into biodiesel (Government of Malta 2007).

Papua New Guinea

Incentives: Papua New Guinea is experiencing great demand for palm oil, particularly from India and China. This is providing economic benefit to the country's palm oil producers. Throughout much of the country, the government supports the expansion of palm oil plantations. In 2002, the government announced several incentives to encourage the increased production of palm oil, including tax credits and decreased import taxes (Blanco 2007).

Production & Markets: Although palm oil is Papua New Guinea's main agricultural export and the country produces so much palm oil that it is the world's third largest exporter (after Indonesia and Malaysia), people on the island of Bouganville have begun producing biodiesel from coconut oil in backyard refineries (Blanco 2007). This occurred following shortages in oil imports, which slowed the country's economy.

Feedstock: Copra (coconut) oil, palm oil

Singapore

Incentives: The country's Economic Development Board has a program named "Spring Startup Enterprise Development Scheme," or Spring Seeds, which will match start-up funding from private investors. Additionally, in 2007, the government's Budget 2007 indicated the country's desire to draw investors via a two percent reduction in the corporate tax rate. Rather than paying twenty percent, the tax rate has been lowered to eighteen percent, which is still slightly higher than Hong Kong (17.5 percent). Additionally, the government decided to increase the corporate partial tax exemption threshold, which will be an effective tax rate of 8.9 percent for 80 percent of the companies located in Singapore (Singapore Government 2007). There is a full exemption of all income taxes for the first S$100,000 (US$65,329) for the first three years for new companies setting up shop in Singapore, thereby bringing the effective tax rate even lower (Singapore Government 2007).

Production & Markets: In 2003, the company Biofuel Research began as a small-scale producer of biodiesel from waste cooking oil. The company began selling biodiesel in 2004 but had a difficult time finding clients because petroleum diesel fuel was inexpensive, so there was little economic incentive for biodiesel use (Lwee 2007). As the cost of petroleum diesel fuel increased and the demand for alternative fuels grew, Biofuel was able to open a production plant in Tuas. The plant has a monthly capacity of 1,500 tons; however, production has been slowed by a shortage in waste cooking oil.

In 2005, two foreign companies announced that they would build Singapore's first large-scale commercial biodiesel plants on Jurong Island, which is the country's hub for petrochemicals. A joint venture between Archer Daniels Midland and Wilmar Holdings resulted in Singapore's second biodiesel facility with annual output of 150,000 metric tons. Meanwhile, German business Peter Cremer built a plant that will produce 200,000 metric tons of biodiesel annually starting in 2007. The company hopes to build two more plants on the island within five years. In 2006, the Australian company Natural Fuel Limited announced its plans to build the world's largest biodiesel facility – with a 1.8-million-metric-ton capacity – in Singapore. This plant is scheduled for completion by the end of 2007 (*Biodiesel Magazine* 2007). Meanwhile, Van Der Horst Biodiesel, a joint venture between Van Der Horst Engineering and the Institute of Environmental Science and Engineering, plans to build the first biodiesel plant in Singapore that produces biodiesel from jatropha oil rather than palm oil. The jatropha plants will be grown on land in China and Cambodia. This plant is expected to have a 200,000 annual metric-ton capacity. The plant will be on Jurong Island, and there are plans for a second facility in Johor (Biopact Team 2007b).

Much of the biodiesel will be produced for export, making Singapore a strategic locale for production because of its maritime connectivity and facilities. Singapore can ship biodiesel around the world with relative ease.

Feedstock: Palm oil from Malaysia and Indonesia, jatropha grown in Cambodia and China

Thailand

Incentives: Thailand launched a national biodiesel program in 2001. This was accomplished by the Royal Chitralada Project, which was created to assist rural farmers. By 2006, several biodiesel plants were in operation. The government expects to mandate B5 use by 2011, requiring the production of nearly 4 million litres of biodiesel daily. The government is involved in several pilot projects.

Production & Markets: Thailand's government hopes that biodiesel production will offset petroleum diesel imports and provide an alternative market for excess agricultural goods. About fifteen filling stations, located in Bangkok and Chiang Mai, sell B5.

Feedstock: Palm oil, coconut oil, waste vegetable oil, jatropha

United Kingdom

Incentives: The United Kingdom's Renewable Transport Fuels Obligation (RTFO) provides incentives for biofuels, including biodiesel. Passed in 2005, the RTFO requires that some domestically sold fuel should come from a renewable energy source: 2.5% volume by 2008, 3.75% by 2009, and 5% by 2010 (Home-Grown Cereals Authority n.d.). The UK also provides a duty rebate of 20 pounds/litre and will continue to do so through September 2009.

Production & Markets: An increasing number of filling stations, including supermarket filling stations, are selling B5 and B100

biodiesel. This biodiesel is being produced by small-scale producers and coops, who sometimes sell their product for several pence/liter less than the cost of petroleum diesel. Some farmers are making biodiesel for use in their farm equipment.

In 2005, the first large-scale biodiesel plant opened in Scotland. It manufactures about 13 million gallons of biodiesel annually. Since then, large-scale producers Tesco and Greenergy have begun making biodiesel as well. As the demand for biodiesel grows, most biodiesel plants are expected to be located in the eastern UK with several in Scotland.

Feedstock: Rapeseed

Zambia

Incentives: There do not appear to be any incentives for biodiesel production in Zambia at this time.

Production & Markets: The National Association for Peasant and Small-Scale Farmers of Zambia (NAPSSF) has announced that, in 2008, approximately 300,000 small-scale farmers will begin raising biofuel crops on over 150,000 hectares of land. This is expected to provide jobs for some of Zambia's large rural population, including underemployed and unemployed farmers. It also will decrease Zambia's dependence on foreign oil. The president of the National Association for Peasant and Small-Scale Farmers of Zambia (NAPSSF) is urging the government to create policy frameworks for the production of biodiesel.

Feedstock: Jatropha is seen as a crop with the potential to alleviate poverty in farming communities.

Appendix B. Biodiesel Cooperatives

Several biodiesel groups start out or consider becoming cooperatives. This section explores this type of business entity. It introduces the concept of the cooperative and discusses cooperative principles. Then, it explains how to start a cooperative and how to become an official, registered entity.

What Is a Cooperative (Co-op)?

The International Cooperative Alliance defines a cooperative (co-op) as "an autonomous association of persons united voluntarily to meet their common economic, social, and cultural needs and aspirations through a jointly-owned and democratically-controlled enterprise" (Wikipedia 2007b). Thus, a cooperative is a business entity that is owned and democratically controlled by a group of members who benefit from the use of its goods and services. Co-ops range in size from small shops to large Fortune 500 companies. In many ways, they are like any other business. They have similar physical facilities, perform similar functions, and must follow sound business practices. Generally, they are incorporated under state law by filing articles of incorporation that grant them the right to do business. The leadership group writes up bylaws and other necessary legal papers. Members elect a board of directors, which sets policies and designates a manager or group to implement them and run the day-to-day operations.

In several important ways, however, co-ops are unique. A co-op's purpose, ownership, control, benefit distribution, and extremely high customer satisfaction are key differences. Large or small, co-ops operate for the benefit of and are governed by their member-owners. In the process, co-ops provide jobs, support business and personal needs, and enhance quality of life. In small rural towns or in big cities, co-ops are business partners, community patrons, and neighbors. They offer members strength in the marketplace and a sense of community. Co-ops can sell consumer goods, such as food, supplies, outdoor equipment, or biodiesel. Co-ops also can provide services, such as electricity, housing, insurance, telecommunications, biodiesel delivery, testing, or marketing. Almost every consumer need can be met by a cooperative.

Not unlike community supported agriculture, community supported energy (CSE) relies on the local community rather than out-of-state corporate entities to develop, site, and own an energy operation, such as a small wind development or a biodiesel production facility. Cooperative ownership of local renewable energy projects offers many benefits, including the creation of new jobs and business opportunities while expanding the tax base, community building, options for cooperative ownership, and minimizing potential conflict between a local community and developers (Pahl 2007b).

Local ownership is the key ingredient that transforms what would otherwise be just another corporate energy project into an engine for local economic development and greater energy security. Community-based energy initiatives can provide at least a basic supply of energy for a decentralized, localized economy. Localized energy initiatives can provide significantly more value over larger, centralized models. For instance, locally generated power can prevent the energy loss that occurs through long-term energy transmission (Pahl 2007a). Furthermore, farmers who sell their crops to large-scale bioenergy producers might experience financial gain, but they are likely to be more successful if they own part of the energy production company. Aside from the benefit to the owner, the community or region benefits economically from localizing profit streams, and local sources for service needs tend to be sourced locally (i.e., legal, accounting, construction). More money remains in a local area to bolster the local economy.

Biodiesel Cooperatives

Biodiesel producers generally fall into three main categories:

1. Large-scale commercial: corporate or community-owned and subject to a wide range of federal, state, and local regulations due to size and customers
2. Small-scale: typically produces between 100,000 to 500,000 gallons per year and is too large to go undetected by government but too small to easily afford the high cost of meeting regulatory requirements for selling ASTM-compliant fuel to the general public
3. Individual: "homebrew" producers who produce for themselves and tend to operate below the radar of most government oversight

Biodiesel co-ops generally fall into one of two categories – large, farmer-owned cooperatives or small community-based cooperatives run by farmers or non-farmers. Historically, large farmer-owned cooperatives have been located primarily in the Midwest and dominated by soy farming. One example is the West Central Co-operative in Iowa, which represents about 3,500 farmer members with a 12-million-gallon capacity biodiesel plant that processes 90 million pounds of soybean oil into biodiesel per year (Pahl 2007a).

Small-scale community based initiatives generally do not sell homemade fuel to the public because of the regulatory hurdles. They generally fall into one of three main categories (Pahl 2007a):

1. Very small-scale producer: involves a couple of experienced fuelmakers who take care of all fuel-making chores for themselves and a small group of members.

2. Bulk-buyer: eliminates production completely (in addition to health and safety hazards associated with biodiesel production) and offers discounted fuel to co-operative members. Some bulk-buyers co-ops have a large, central storage tank where a load is delivered and everyone fills up at. Others have a distributed model with several smaller storage tanks in multiple locations.

3. Large volunteer producer group: Provides the opportunity for worker members to get trained to make fuel on a volunteer basis.

Case Study: New River Valley Biodiesel Co-operative, Inc., Blacksburg, Virginia

The New River Valley Biodiesel Co-operative, Inc., began in Fall 2006 in order "to purchase high quality biodiesel as a group, thus gaining the benefit of having a centralized biodiesel tank and pump and the ability to buy in volume gaining the benefits of delivery and lower prices." The co-op comprises a group of interested biodiesel users who are working to make biodiesel more accessible and available in the New River Valley area of Virginia. Many of the co-op's members use biodiesel individually and/or promote the use of biodiesel. Currently, there is no readily accessible source of biodiesel in the region. In fact, biodiesel users have to drive to the Virginia Biodiesel Refinery in Richmond, Virginia, to fill their drums for transport back to the New River Valley area.

Cristina Siegel is one of the founding members and serves as Board Member and co-op member. Cristina became involved because she had gotten "so many questions about my personal usage that I realized we needed to form some kind of network to get everyone together and share information." So, she began a biodiesel listserve, which grew quickly. Using the list serve as a springboard, interested biodiesel users started meeting to brainstorm ideas on how to make, get, and use biodiesel.

Members believed that the fastest and easiest way to have a readily available source of biodiesel was to join together to locate a large size tank and pump to which a distributor would deliver. Cristina says, "We found a distributor in Rocky Mount who would deliver if we had minimum 500-gallon orders. This arrangement allowed folks to buy in bulk, get a better price, and have it delivered locally." The group is interested in building up the area's biodiesel infrastructure by, for example, finding a fuel retailer that would install a tank. Some members also would like to make their own biodiesel.

Although Cristina had purchased one of the tanks and the pump setup with a $500 grant received from the Virginia Soybean Association in 2004, the Co-operative did not receive any funding to support its mission. She donated the tank and pump to the organization.

Cristina recalls that determining which regulations pertained to biodiesel and to them was a challenge. She says, "Advice was mixed and often conflicting." One of their founding members contacted the Virginia Department of Motor Vehicles (DMV), local planning and zoning boards, the Virginia Department of Environmental Quality (DEQ) and others. She remembers, "Although DEQ had no specific regulations relating to biodiesel, they generally gave us advice based on us using petrodiesel, such as spill containment, which often seemed inappropriate for biodiesel. We still don't have a completely clear understanding, but believe that we have done everything possible to be legal and safe."

To handle all of the legal issues related to fuel resale, storage, and so forth, the group decided that the co-operative structure would be the best group structure. The legal arrangement also provides protection for group members and the landowner on whose land the tanks were located. The size of its tank is small enough to avoid regulatory issues with the DEQ. The co-op does not sell fuel and, thereby, avoids taxation issues. It pays its road taxes up front to the distributor.

In June 2007, it was incorporated as a member-owned co-operative. Following incorporation, the New River Valley Biodiesel Co-operative placed its first 500-gallon order of biodiesel. The fuel tank is filled every one to two months. The co-op's tank is located on a member's farm on the outskirts of Blacksburg, Virginia. Co-op members fill up their cars directly, fill 55-gallon drums at fuel delivery time, or fill 5-gallon jugs from the tanks. Members pay for their anticipated two-month fuel need in advance. Cristina explains, "We have a very simple honor system by which users write down on a log located at the tanks the amount of fuel they've pumped at each visit." The fuel manager periodically collects that information and keeps track of each person's fuel usage related to the amount they paid for originally.

Based on the knowledge of experienced users, the co-op's first decision was to always purchase and use only ASTM-certified fuel. They set up two-275 gallon linked tanks for storage. These are kept outside but are somewhat sheltered. The tanks are tilted slightly toward a drain valve on the bottom so that water contamination can be drained off periodically. They organization uses a triple filter system through which all fuel is pumped. The filters include a screen particulate filter, a 15-micron water blocking filter and a 10mm filter. "We hope to avoid long-term biodiesel storage issues by using all the fuel in the tanks on a monthly or bimonthly basis," Cristina reports. The co-op uses 100% biodiesel and is currently working to develop a plan for winter storage that may include additives and a small tank heater.

Finding a distributor for the fuel was a challenge. Cristina says that her organization wanted to buy B100 in 500-gallon increments. Most distributors had large trucks and wanted to deliver at least 2,000 gallons at a time. Furthermore, some distributors were very leery of selling B100 to folks who may not know the properties of and issues storing the fuel."

To ensure that it can meet its members' biodiesel needs, it has kept its membership numbers down. Currently, there are fifteen members. Before allowing new members, Cristina anticipates buying a second tank and/or relocating to increase capacity. She says, "We do want to grow and be able to take on more members but are cautious to really get the current system up and running before we grow too much more."

When asked about areas of improvement, Cristina explained that the members were so focused on ordering biodiesel as quickly as possible that they did not think about the day-to-day operations, equipment maintenance, and fuel orders as much as they should have. She states, "As issues came up, such as a malfunctioning pump, it became clear that we had not clearly established how the group would function." Some people had been doing much of the work while others had bought fuel regularly but did not volunteer much of their time toward the group's daily functions.

Addressing this was key to the proper functioning of the organization. At a meeting, the Board emphasized that the co-op is a member-operated group and that part of the membership agreement involves volunteering for activities that keep the co-op running. To that end, the group established the following duties for its volunteers:

- Officers – Chair, Vice-Chair, etc.
- Fuel Ordering – Work with wholesalers, negotiate discounts, schedule fillups
- General Site Manager – Check fuel levels, check for problems
- Repairs – Maintain pump, diagnose problems, repair problems when possible, call for help as needed
- Back-up repairman – Secondary handyman
- Recordkeeper – Keep track of fuel orders and fuel use

- Membership Committee – Recruit members, maintain agreements
  - Webmaster – Create/maintain web site
  - Outreach – Work with town, retail pumps, public
  - Fuel Specialist – Answer questions about fuel quality/ use

Once it has worked out these issues, Cristina expects that the group will be able to focus on creating an informative web site and to promote biodiesel through active outreach efforts.

Co-op Principles

Worldwide, cooperatives tend to operate under the same principles. According to the National Cooperative Business Association, "cooperatives are based on the values of self-help, self-responsibility, democracy, equality, equity, and solidarity" (National Cooperative Business Association n.d.(b)). Their members believe in the "ethical values of honesty, openness, social responsibility, and caring for others" (National Cooperative Business Association n.d.(b)). The International Cooperative Alliance has developed seven principles for co-ops, and those principles are shown below.

"*Voluntary and Open Membership*. Co-operatives are voluntary organisations, open to all persons able to use their services and willing to accept the responsibilities of membership, without gender, social, racial, political or religious discrimination.

Democratic Member Control. Co-operatives are democratic organisations controlled by their members, who actively participate in setting their policies and making decisions. Men and women serving as elected representatives are accountable to the membership. In primary co-operatives members have equal voting rights (one member, one vote) and co-operatives at other levels are also organised in a democratic manner.

Member Economic Participation. Members contribute equitably to, and democratically control, the capital of their co-operative. At least part of that capital is usually the common property of the co-operative. Members usually receive limited compensation, if any, on capital subscribed as a condition of membership. Members allocate surpluses for any or all of the following purposes: developing their co-operative, possibly by setting up reserves, part of which at least would be indivisible; benefiting members in proportion to their transactions with the co-operative; and supporting other activities approved by the membership.

Autonomy and Independence. Co-operatives are autonomous, self-help organisations controlled by their members. If they enter to agreements with other organisations, including governments, or raise capital from external sources, they do so on terms that ensure democratic control by their members and maintain their co-operative autonomy.

Education, Training and Information. Co-operatives provide education and training for their members, elected representatives, managers, and employees so they can contribute effectively to the development of their co-operatives. They inform the general public – particularly young people and opinion leaders – about the nature and benefits of co-operation.

Cooperation among Cooperatives. Co-operatives serve their members most effectively and strengthen the co-operative movement by working together through local, national, regional, and international structures.

Concern for Community. Co-operatives work for the sustainable development of their communities through policies approved by their members" (International Co-operative Alliance 2007).

Starting a Biodiesel Cooperative or Group

Starting a cooperative is a complex undertaking. The sooner you gather a group of interested individuals the better. Bumper stickers, newsletters, web sites, meetings, and workshops are all great ways to build awareness and bring attention to your developing group. Online forums are an excellent starting place to gather ideas, build awareness of your vision, and attract like-minded individuals willing to join forces and contribute to the cause.

The founding members hold the biggest responsibility for starting a fully functional cooperative. Once you have a small, dedicated group of visionaries, start discussing the needs and challenges facing your area with respect to biodiesel. Common issues facing communities tend to be availability of B100 and biodiesel blends, support and awareness, proper education of homebrew technologies and legal and safe operation, and fuel quality. Identify ways that a cooperative could combat the issues and challenges revealed during initial brainstorming sessions and meetings.

A new cooperative will have a better chance of surviving if its model is simple and begins with only a couple of projects. Leave larger goals and expanded territories for the future when the co-op is experienced and more established.

It is advisable to conduct a feasibility study for the proposed cooperative to determine if your model is feasible for your community and if you have overlooked any key details. For example, ask yourselves: How many people are committed to becoming members? How many are needed? What will the membership dues schedule be? A start-up requires sufficient numbers of potential members with enough raised capital to make the business viable unless members intend on volunteering large amounts of their time and money. If you have decided that your group is going to increase the availability of biodiesel in your area, find out in what quantities and at what price distributors will sell biodiesel. Even if your group is going to focus on biodiesel production it is still recommended to buy commercial biodiesel during the initial stages of co-op operation.

Keep in mind that you are forming a cooperative with cooperation being the primary goal. Attract like-minded individuals who will assist in moving the original vision forward. The functionality of your group is dependent on the healthy relationships of those that run the operation. Listen, try to stay open-minded, and be positive.

The commitment and willingness of members to finance and patronize the new venture is crucial. Co-ops must have ownership capital to conduct day-to-day operations, to provide the necessary facilities, and to create a base for obtaining external financing. Member capital is collected mainly through memberships.

New cooperatives may be able to obtain private or government loans, grants, or guarantees depending on what programs are currently available and the specific problem that the co-op intends to solve. In seeking financing, leadership should develop a business plan, which describes the need for the co-op, membership support, marketing plans, projected cash flow, and operational results for the next three to five years.

Assigning an experienced, qualified management team is one of the most crucial tasks facing a growing co-op. Success is more dependent on the leadership than on almost any other factor. It is the responsibility of the board of directors to ensure that quality managing and decisions are being made by the management group. The management group is responsible for carrying out the policies of the board. Good board/management relationships are essential to the effectiveness of a cooperative. Every individual plays an important role in a co-op, and responsibilities should be clearly defined, mutually understood, and respected.

The NCBA Cooperative Business Journal and USDA's magazine for cooperative business printed tips for co-op success and can be found on the NCBA's web site: http://www.ncba.coop/ab-coop_ab_success.cfm. Forming an official and legal co-op is a large undertaking and should include responsible decision-making and a respect for co-op members and the original principles, mission, and vision of the organization.

Becoming an Official Cooperative

Assuming that your co-op's leadership group determines that the co-op model is feasible and will bring value to your community, the formal process of becoming a legal entity begins. Steps include incorporating the entity, increasing membership, developing a business plan, and securing financing. As with any start-up, advice may be necessary from time to time from a variety of people experienced in cooperative organization, business development, legal issues, financing, and accounting. Legal help is important. Lawyers can prepare the corporate documents and give advice on legal compliance, securities, and tax laws.

According to the National Cooperative Bank's and National Cooperative Business Association's book *How to Organize a Cooperative*, the following steps can be used in the formation of a cooperative:

- Hold an initial meeting with founders to brainstorm how a biodiesel cooperative would provide value to your community (this is a crucial step, provide an ample amount of time and energy to discuss if an actual need exists).

• Hold an exploratory community meeting and determine if a biodiesel cooperative is necessary. Elect a steering committee to begin discussing the intricacies if a consensus is reached to continue.

• Conduct an initial feasibility study (does not need to include complex financial analysis yet)

• Hold another community meeting, discuss feasibility results, and vote on whether to continue.

• Conduct a market or supply and cost analysis

• Hold another community meeting to discuss results of the analysis, and vote on whether to continue.

• Conduct a financial analysis and develop a business plan

• Hold another community meeting to discuss results of financial analysis and business plan. Vote on whether to keep the steering committee intact or make changes.

• Draw up legal papers necessary to incorporate

• Hold a meeting with all potential charter members, adopt bylaws, and elect a board of directors.

• Hold an initial meeting of the board of directors, elect officers, and determine and delegate responsibilities to implement business plan

• Conduct a membership drive (be creative!)

• Acquire capital (can include loans, grants, or donations)

• Hire a cooperative manager

• Find adequate operation facilities that meet the needs of your cooperative

• Begin operation (National Cooperative Business Association. n.d.(a))

The articles of incorporation and the bylaws are the main organizational documents for your cooperative. Other documents, such as contracts between the co-op and its members (i.e., membership application and certificate), might be required.

The articles of incorporation typically contain the following components and state what type your cooperative is and its function:

- "Name of the cooperative
- Principle place of business
- Purposes and powers of the cooperative
- Proposed duration of the cooperative
- Names of the incorporators
- A provision for redemption of member equity (may be in the bylaws)" (National Cooperative Business Association. n.d.(a))

The bylaws typically contain the following components and express the rights and duties of the board and the co-op members as well as addressing the day-to-day operations:

- "Requirements for membership
- Rights and responsibility of members
- Grounds and procedures for member expulsion
- Procedures for calling and conducting membership meetings
- Voting procedures
- Procedures to elect or remove directors and officers
- The number, duties, terms of office, and compensation of directors and officers
- Time and place of the directors' meetings
- Dates of the fiscal year
- Information on how the net earnings will be distributed
- Other rules for management of the cooperative" (National Cooperative Business Association. n.d.(a))

Other Organizational Options

Clearly, there are several options for biodiesel production. Tom Leue, of Yellow Biodiesel, suggested that no one try to go it alone because collection, production, and marketing are simply too much for one person.

Kumar Plocher, of Yokayo Biofuels, considered forming a cooperative before settling on an "S" corporation. He felt that, despite the potential enthusiasm of co-op members, the skills necessary for running Yokayo Biofuels might not necessarily be reflected in a cooperative's membership. Recognizing the need to deal with regulatory issues and administrative tasks, he chose the corporation as the most appropriate form for his group.

For those who are just learning the biodiesel ropes and feel a bit overwhelmed by starting a co-op or a company, small groups of two to ten people usually can work together without the official registration and formalities of a co-op or company. The group will need to organizing meetings to determine how much it wants to produce, what the regulations are, what permitting is necessary, and so forth.

Of course, the group does not have to produce biodiesel – it can be a purchasing group that receives delivery of 55-gallon or 250-gallon totes of B100. Or it could focus on promoting and increasing awareness of biodiesel or biofuels.

References

ABG Biodiesel. "Markets." *ABG Biodiesel* (2007). Available from http://www.abgbiodiesel.com/Australia/ab_markets.asp (cited February 18, 2007).

AFX News Limited, 2006. "Japan targets 40 pct of cars to use biofuels by yr to March 2009 – report". *Forbes* (June 20, 2006). Available from http://www.forbes.com/business/feeds/afx/2006/06/20/afx2826178.html (cited June 29, 2007).

Alabama Farmers Federation. "Wheat and Feed Grains: Did you know?" (n.d.). Available from http://www.alfafarmers.org/commodities/grains.phtml (cited September 6, 2006).

Alberta Energy and Utilities Board. "Alberta's Reserves 2003 and Supply/Demand Outlook 2004-2013." (June 2004). Available from http://www.eub.gov.ab.ca/bbs/products/STs/st98-2004.pdf (cited July 29, 2007).

Alternative Fuels Data Center. "Physical and chemical properties of gasoline." US Department of Energy (n.d.). Available from http://www.eere.energy.gov/afdc/pdfs/fueltable.pdf (cited September 27, 2006).

American Soybean Association. "U.S. Soybeans." (2005). Available from http://www.asajapan.org/english/us_soybeans/miracle.html (cited August 8, 2007).

Andreas, Dave. "Collaborative Biodiesel Tutorial: Storage Considerations." (2005). Available from http://www.biodieselcommunity.org/storageconsiderations/ (cited June 13, 2007).

Avendaño, Tatiana Roa. "Colombia: Biodiesel from oil palm" *World Rainforest Bulletin* 112 (November 2006). Available from http://www.wrm.org.uy/bulletin/112/Colombia.html (cited July 27, 2007).

Bartlett, Albert. "Thoughts on Long-Term Energy Supplies: Scientists and the Silent Lie." *Physics Today* 57 (2004): 53-55.

Batey, J. "Interim report of test results." *Massachusetts Oilheat Council Biodiesel Project*. Massachusetts Oilheat Council (2002).

BBC News. "UN warns on impacts of biofuels." BBC News (May 9, 2007). Available from http://news.bbc.co.uk/2/hi/science/nature/6636467.stm (cited June 21, 2007).

Bently Tribology Services. "Biodiesel Testing." (2007). Available from http://www.biodieseltesting.com/tests.php (cited July 29, 2007).

Biodiesel Magazine. "Natural Fuel starts construction of Singapore biodiesel plant". *Biodiesel Magazine* (January 2007). Available from http://www.biodieselmagazine.com/article.jsp?article_id=1340 (cited June 29, 2007).

Biopact Team. "Swiss technology institute launches 'Roundtable on Sustainable Biofuels.' " (2007). Available from http://biopact.com/2007/04/swiss-technology-institute-launches.html (cited September 4, 2007).

_____. "Singapore's first jatropha biodiesel plants eye exports to China." Biopact Team (March 22, 2007). Available from http://biopact.com/2007/03/singapores-first-jatropha-biodiesel.html (cited June 29, 2007).

_____. "Zambia: 300,000 small-scale farmers to grow energy crops to generate incomes, jobs, alleviate poverty." (2006). Available from http://biopact.com/2006/11/zambia-300000-small-scale-farmers-to.html (cited February 16, 2007).

Blach, Oliver, and Rory Carroll. "Massacres and paramilitary land seizures behind the biofuel revolution." *The Guardian* (2007).

Blades, T., M. Rudloff, and O. Schulze. "Sustainable SunFuel from CHOREN's Carbo-V Process." Paper read at International Symposium on Alcohol Fuels and Other Renewables (September 26-28, 2005) at San Diego, California.

Blanco, Sebastian. "Coconut oil powers vehicle, generators in Bougainville, Papua New Guinea." *AugoblogGreen* (May 14, 2007). Available from http://www.autobloggreen.com/2007/05/14/coconut-oil-powers-vehicle-generators-in-bougainville-papua-ne/ (cited June 29 2007).

Briggs, Michael, Joseph Pearson, and Ihab Farag. "Biodiesel Processing." University of New Hampshire (n.d.). Available from http://www.unh.edu/p2/biodiesel/media/nhsta43.doc (cited July 29, 2007).

British Petroleum Co. "BP Statistical Review of World Energy 2005." London: British Petroleum Co.

———. "Statistical Review of World Energy 2007." British Petroleum (2007). Available from http://www.bp.com/productlanding.do?categoryId=6848&contentId=7033471 (cited July 29, 2007).

Brodrick, Christie-Joy, et al. "Farm-Scale Biodiesel Process Variations: An Analysis of Equipment Operations of Four Manual Small-Scale Batch Processors." Unpublished (2005).

Bryan, M. "The economics of ethanol and bio-diesel production." USDA Agricultural Outlook Forum - 2002 (2002). Available from http://www.usda.gov/oce/waob/Archives/2002/speeches/bryanppt.pdf (cited November 18, 2004).

Bush, George W. State of the Union Address. White House, January 2006. Available from http://www.whitehouse.gov/stateofthe-union/2006/index.html (cited September 27, 2006).

Campbell, C.J., and J.H. Laherrere. "The End of Cheap Oil." *Scientific American*, March 1998.

Canadian Renewable Fuels Association. "Biodiesel Around the World." (March 30, 2006). Available from http://www.greenfuels.org/biodiesel/world.htm (cited June 29, 2007).

Canakci, M., and J. Van Gerpen. "Biodiesel production from oils and fats with high free fatty acids." *ASAE* 44 (6) (2001):1429-1436.

Center for the Study of Carbon Dioxide and Global Change. "Climate Models: Approximations and Limits." *CO2 Science* (2006). Available from http://www.co2science.org/scripts/CO2Science-B2C/subject/other/climate_models.jsp (cited July 29, 2007).

Chin, Loh Kim. "Singapore to host two biodiesel plants, investments total over S$80m." October 26, 2005.

City of Longmont. "The City's Grease Problem." (2007). Available from http://www.ci.longmont.co.us/pwwu/ipp/FOG.htm (cited August 8 2007).

Clean Air Initiative for Asian Cities. "Malaysia Wants to be World's Biggest Biodiesel Producer." Clean Air Initiative for Asian Cities (January 30, 2007). Available from http://www.cleanairnet.org/caiasia/1412/article-70344.html (cited June 29 2007).

Climate Ark. "Action Alert: Indonesia's Biofuel Expansion on Rainforest Peatlands to Accelerate Climate Change." (February 18, 2007). Available from http://www.climateark.org/alerts/send.asp?id=indonesia_peatland (cited June 29, 2007).

Crawford, Peter. "The Exciting Subject of Grease Traps. Green Hotels in the Green Mountain State." (n.d.). Available from http://www.vtgreenhotels.org/articles/greasetrap.htm (cited July 29, 2007).

Commission of the European Communities. "Biofuels Progress Report: Report on the progress made in the use of biofuels and other renewable fuels in the Member States of the European Union." Brussels: Commission of the European Communities (2007).

Commission of the European Communities. "Renewable Energy Road Map: Renewable energies in the 21st century: building a more sustainable future." Brussels: Commission of the European Communities (2006).

de Sousa, Paulo Teixeira. "The Ethanol and Biodiesel Programmes in Brazil: A Brief Discussion." (n.d.). Available from http://www.

intech.unu.edu/events/workshops/hfc05/abstracts.php (cited February 16, 2007).

Defense Energy Support Center. *Fact Book FY 2005* (28th), 2005. Available from http://www.desc.dla.mil/DCM/Files/Fact05Revised.pdf (cited July 29, 2007).

Deffeyes, Kenneth S. *Hubert's Peak: The Impending World Oil Shortage.* Princeton, NJ: Princeton University Press, 2001.

_____. *Beyond Oil: The View from Hubbert's Peak.* New York, NY: Farrar, Straus and Giroux, 2005.

Dietzen, Jolie. "Opposing Palm Oil Plantations." (October 23, 2006). Available from http://www.greengrants.org/grantstories.php?news_id=104 (cited June 29, 2007).

Dunn, Seth. "Hydrogen Futures: Toward a Sustainable Energy System." *International Journal of Hydrogen Energy* (2002). Available from http://www.sciencedirect.com/science?_ob=ArticleURL&_udi=B6V3F-44TSXJS-1&_coverDate=03%2F31%2F2002&_alid=423304109&_rdoc=1&_fmt=&_orig=search&_qd=1&_cdi=5729&_sort=d&view=c&_acct=C000035098&_version=1&_urlVersion=0&_userid=650596&md5=2f5f1451b7013c14da6134099a2b408e#bib73 (cited July 11, 2006).

DuPont. "BioButanol Fact Sheets." DuPont (2006). Available from http://www2.dupont.com/Biofuels/en_US/facts/index.html (cited July 29, 2007).

Dyni, J.R. "Geology and Resources of Some World Oil-Shale Deposits." *Oil Shale* 20(3) (2003):193-252.

Eastern Connecticut State University. "Basics on restaurant grease generation." Regional workshop on biodiesel for New England (March 26, 2003). Available from http://www.easternct.edu/depts/sustainenergy/calendar/biodiesel/Gridley%20-%20BASICS%20ON%20RESTAURANT%20GREASE%20GENERATION%20.pdf (cited July 29, 2007).

Edible Oil Refining Corporation. "Biodiesel." (2007). Available from http://www.eorc.com.mt/biodiesel.html (cited July 30, 2007).

Encyclopedia Britannica. "Cellulose." *Encyclopedia Britannica*, n.d.. Available from http://www.britannica.com/eb/article-9022028/ cellulose (cited September 27, 2006).

ENVIRON Corporation. "Environmental effects of releases of animal fats and vegetable oils of waterways." Arlington, Virginia. (1993).

Environmental Energy Inc. "Butanol Replaces Gasoline." ButylFuel, LLC (2006). Available from http://www.butanol.com/index.html (cited July 29, 2007).

Environmental Protection Agency. "Grease trap waste management." Queensland Government (August 2006). Available from http://www.epa.qld.gov.au/publications/p01365aa.pdf/Grease_trap_waste_management.pdf (cited July 29, 2007).

Environmental Protection Department. "Grease and Oil Wastes - Problems and Solutions." The Government of Hong Kong (April 26, 2006). Available from http://www.epd.gov.hk/epd/english/environmentinhk/water/guide_ref/guide_wpc_gt_1.html (cited July 29, 2007).

Estill, Lyle. *Biodiesel Power: The Passion, the People, and the Politics of the Next Renewable Fuel.* Gabriola Island, BC, Canada: New Society Publishers, 2005.

_____. "Getting Legal (in Piedmont Biofuels Energy Blog)." (December 22, 2006). Available from http://energy.biofuels.coop/general/2006/12/22/getting-legal/ (cited June 23 2007).

_____. "The Permits." In Piedmont Biofuels Energy Blog (January 5, 2007). Available from http://energy.biofuels.coop/general/2007/01/05/the-permits (cited June 23, 2007).

_____. Personal communication, June 1, 2007.

_____. "Sustainble Biodiesel Alliance" (June 29, 2007). Available from http://energy.biofuels.coop/general/2007/06/29/sustainable-biodiesel-alliance/ (cited July 29, 2007).

Fitzgerald, Michael. "India's Big Plans for Biodiesel: Researchers are developing new methods for cultivating a plant called jatropha." *MIT Technology Review* (December 27, 2006). Available from http://www.technologyreview.com/Energy/17940/page1/ (cited April 6, 2007).

Ford. "Biodiesel Technology." (May 27, 2005). Available from https://www.fleet.ford.com/showroom/environmental_vehicles/ BiodieselTechnology.asp (cited July 30, 2007).

Foreign Agricultural Service. "World Agricultural Production." US Department of Agriculture (December 3, 2003). Available from http://www.fas.usda.gov/wap/circular/2003/03-02/tables.html (cited July 29, 2007).

Freckmann, Chad. "Small-Scale Production and Use." (April 3, 2006). Available from http://www.cisat.jmu.edu/biodiesel/presentations06/BRCFI-U%20Biodiesel%20Conf.ppt (cited April 26, 2006).

Genecor International. "Genencor Offers Industry Transforming Technology for Ethanol Production." (n.d.). Available from http://www.genencor.com/wt/gcor/ethanol (cited September 27, 2006).

Georgia Department of Natural Resources. "Managing fats, oils, and grease in wastewater." Pollution Prevention Assistance Division (2007). Available from http://www.p2ad.org/documents/ ci_pubs_fog.html (cited July 29, 2007).

Gerpen, Jon Van. "Biodiesel Production and Fuel Quality." University of Idaho (2005). Available from http://www.uidaho.edu/ bioenergy/biodieselED/publication/01.pdf (cited 2006).

Government of Malta. "Increase registered on biodiesel consumption." (November 15, 2006). Available from http://www.mrae. gov.mt/pressrelease.asp?id=1239 (cited July 27 2007).

Green Car Congress. "National Express Group Suspends First-Generation Biodiesel Trials." (August 6, 2007). Available from http://www.greencarcongress.com/2007/08/national-expres. html#more (cited September 4, 2007).

Green Car Congress. "Fortum Building Biodiesel Plant in Finland." (February 15, 2005). Available from http://www.greencar-congress.com/2005/02/fortum_build_bi.html (cited July 30, 2007).

_____. "Indian President Calls for 60M Tonnes of Biodiesel/Year by 2030; 50% of Current Oil Consumption." Green Car Congress (June 9, 2006). Available from http://www.greencarcongress. com/2006/06/indian_presiden.html (cited July 27, 2007).

Green Oasis Environmental Incorporation. "The Environmental Impact." (2002). Available from http://www.greenoasis.com/ environment/ (cited May 2004).

Greene, David. *Transportation and Energy*. Lansdowne, Virginia: Eno Transportation Foundation, Inc., 1996.

Greene, David L., and Andreas Schafer. "Reducing greenhouse gas emissions from U.S. transportation." PEW Center on Global Climate Change (May 2003). Available from http://www.envi-rolink.org/resource.html?itemid=200305291231210.235971 &catid=6 (cited July 29, 2007).

Gutro, Rob. "2005 Warmest Year in Over a Century." National Aeronautics and Space Administration (January 24, 2006). Available from http://www.nasa.gov/vision/earth/environment/2005_warm-est.html (cited July 29, 2007).

Hagen, K.M. "Taking a Tax Credit for Alcohol, Biodiesel, and Agri-biodiesel Fuels" (September 26, 2005). Available from http://www.googobits.com/articles/2783-taking-a-tax-credit-for-alcohol-biodiesel-and-agribiodiesel-fuels.htm (cited June 28, 2007).

Hansen, J., R. Ruedy, M. Sato, and K. Lo. "GISS Surface Tempera-ture Analysis, Global Temperature Trends: 2005 Summation." NASA Goddard Institute for Space Studies (January 12, 2006).

Available from http://data.giss.nasa.gov/gistemp/2005/ (cited July 29, 2007).

Hart, Michael G. "Locomotive Emissions Reduction Project." Sierra Railroad Company (n.d.). Available from http://www.sierrarailroad.com/powertrain/loc_emissision.pdf (cited July 29, 2007).

Haumann, B.F. "Renderers give new life to waste restaurant fats." *Inform* 1(8) (1990):722-725.

Hearn, Kelly. "Bio for All: A biodiesel entrepreneur in Argentina spreads seeds of wisdom." Grist (2006).

Home-Grown Cereals Authority. "UK Biofuel Situation." HGCA (n.d.). Available from http://hgca.com/publications/documents/UK_Biofuel_situation.pdf (cited July 30, 2007).

Houghton, J.T., Y. Ding, D.J. Griggs, N. Noguer, P.J. van der Linden, D. Xiaosu, K. Maskell, and C. A. Johnson. *Climate Change 2001: A Scientific Basis.* Cambridge, U.K.: Cambridge University Press, 2001.

Houghton, John T., L. Gylvan Meira Filho, David J. Griggs, and eds. Kathy Maskell. "An introduction to simple climate models used in the IPCC second assessment report." Intergovernmental Panel on Climate Change (February 1997). Available from http://www.ipcc.ch/pub/IPCCTP.II(E).pdf (cited July 29, 2007).

Idso, Sherwood B., Craig D. Idso, and Keith E. Idso. "Enhanced or Impaired? Human Health in a CO2-Enriched Warmer World." Center for the Study of Carbon Dioxide and Global Change (2003). Available from http://www.co2science.org/scripts/Template/0_CO-2ScienceB2C/pdf/health2pps.pdf (cited July 29, 2007).

IFQC. "Fuel Quality in Asia - 2007 Regional Overview and Outlook presentation." (May 2007). Available from http://www.ifqcbiofuels.org/member/PDFs/IFQCAsiaWebinarMay07.pdf (cited June 29, 2007).

Intergovernmental Panel on Climate Change, Working Group III. "Technical Summary." (2007). Available from http://www.mnp.

nl/ipcc/pages_media/FAR4docs/chapters/TS_WGIII_220607.pdf (cited July 29, 2007).

International Code Council. "Topic: Biodiesel Ord." (June 18, 2007). Available from http://www.iccsafe.org/cgi-bin/ultimatebb.cgi?ubb=get_topic;f=1;t=001374 (cited July 29, 2007).

International Co-operative Alliance. "Statement on the Co-operative Identity." (May 26, 2007). Available from http://www.ica.coop/coop/principles.html (cited July 27, 2007).

International Finance Corporation. "Environmental, Health and Safety Guidelines for Oleochemicals Manufacturing." (April 30, 2007). Available from http://www.ifc.org/ifcext/policyreview.nsf/AttachmentsByTitle/EHS_Draft_OleochemicalsMfg/$FILE/IFC+Draft+-+Oleochemicals+Manufacturing+-+Feb+5+2007.pdf (cited July 29, 2007).

Irani. "Australia's SARDI to Produce Bio-diesel from Micro-algae." (July 21, 2006). Available from http://www.ecofriend.org/entry/australias-sardi-to-produce-bio-diesel-from-micro-algae/ (cited February 18, 2007).

International Institute for Sustainable Development (IISD). "Highlights from Wednesday, 1 August." Paper read at First High-Level Biofuels Seminar in Africa, July 30-August 1, 2007, at Addis Ababa, Ethiopia.

Jensen, Mike. "Electricity could take different track." *Union Democrat*, 2002.

John Deere. "Biodiesel Fuel in John Deere Tractors." (February 25, 2005). Available from http://www.biodiesel.org/pdf_files/OEM%20Statements/2004_OEM_john_deere.pdf (cited July 30, 2007).

Johnston, Cathie. "Bio-diesel Basics and Quality Standards." (2006). Available from http://www.cisat.jmu.edu/biodiesel/presentations06/INTERTEK%20Cathie%20Johns%7E0006.ppt (cited April 26, 2006).

Jones, W. "Hydrogen on Track: Trains and Industrial Equipment Now, Cars Later." IEEE Spectrum (August 2006). Available from http://www.spectrum.ieee.org/aug06/4255 (cited September 27, 2006).

Kaihla, Paul. "Soybeans that give you gas." *Business 2.0 Magazine* (August 1, 2006). Available from http://money.cnn.com/2006/07/31/magazines/business2/Soybeans_gas.biz2/index.htm (cited July 27, 2007).

Kao, Ikuko. "Japan Sees Biodiesel Boost With New Fuel Standards." *Reuters News Service* (August 18, 2006). Available from http://www.planetark.org/dailynewsstory.cfm/newsid/37719/story.htm (cited June 29, 2007).

Kaptur, Congresswoman Marcy. "Presentation." Ft. Lauderdale, FL (2005).

Keeling, C.D., and T.P. Whorf. "Atmospheric CO2 records from sites in the SIO air sampling network." In *Trends: A Compendium of Data on Global Change.* Carbon Dioxide Information Analysis Center. Oak Ridge, TN: Oak Ridge National Laboratory, 2005.

Kemp, William H. *Biodiesel Basics and Beyond: A comprehensive Guide to Production and Use for the Home and Farm.* Tamworth, Ontario, Canada: Aztext Press, 2006.

Kendell, James M. "Measures of oil import dependence." U.S. Energy Information Administration (July 22, 1998). Available from http://www.eia.doe.gov/oiaf/archive/issues98/oimport.html (cited July 29, 2007).

Krishna, C.R. "Biodiesel blends in space heating equipment (NREL/SR-510-33579)." Brookhaven National Laboratory (October 2003).

Krukowska, Ewa. "Analysis - Polish Biodiesel Output seen Surging on Law Change." (March 14, 2006). Available from http://www.planetark.com/dailynewsstory.cfm/newsid/35623/story.htm (cited April 21, 2007).

Leidel Energy Serivces. "Michigan Biofuel Production Cooperative Development Plan." (December 2006). Available from http://www.biodieselmichigan.com/Biodiesel_COOP_Plans.pdf (cited May 20, 2007).

Lwee, Melissa. "Out of the frying pan." Singapore Press Holdings Limited (July 27, 2007). Available from http://www.asia1.com/asia1portal/2006/IA/bluesky_IA060802/Blue%20Sky_files/Blue%20Sky%20-%2005_Out%20of%20the%20frying%20pan%20.%20.%20.htm (cited July 29, 2007).

Management Information Services. "Federal incentives for the energy industries." Washington, D.C., 1998.

Mansell, G., R.E. Morris, and G. Wilson. "Impact of biodiesel fuels on air quality and human health." National Renewable Energy Laboratory (May 2003). Available from http://www.eere.energy.gov/afdc/pdfs/33796.pdf (cited July 29, 2007).

Marcus Colchester, Norman Jiwan, Andiko, Martua Sirait, Asep Yunan Firdaus, A. Surambo, and Herbert Pane. *Promised Land: Palm Oil and Land Acquisition in Indonesia - Implications for Local Communities and Indigenous Peoples.* Morton-in-Marsh, England: Forest Peoples Programme and Perkumpulan Sawit Watch, 2006.

Marcus, J. "Iran Enrichment: A Chinese Puzzle?" BBC News (May 18, 2006). Available from http://news.bbc.co.uk/2/hi/middle_east/4995350.stm (cited September 27, 2006).

McCormick, R.L., T.L. Alleman, M. Ratcliff, L. Moens, and R. Lawrence. "Survey of the Quality and Stability of Biodiesel and Biodiesel Blends in the United States in 2004: National Renewable Energy Laboratory." (2005).

McCormick, R.L., J.R. Alvarez, and M.S. Graboski. "NOx solutions for biodiesel." National Renewable Energy Laboratory (February 2003). Available from http://www.nrel.gov/docs/fy03osti/31465.pdf (cited July 29, 2007).

McHugh, D.J. "A guide to the seaweed industry: FAO." (2003).

Methanex. "Methanol Safe Handling and Storage - presentation." Methanex (n.d.). Available from http://www.eere.energy.gov/de/pdfs/road_shows/eugene_methanol.pdf (cited January 18, 2007).

Methanol Institute. "Frequently Asked Questions About the Safe Handling and Use of Methanol." (2006). Available from http://www.biodiesel.org/resources/reportsdatabase/reports/gen/20060401_GEN-370.pdf (cited January 18, 2007).

Moore, Todd. NHDES Air Resources Division, Personal communication, August 21, 2007.

Nakanishi, Nao. "Scientists warn on biofuels as palm oil price jumps." Reuters (May 31, 2007). Available from http://www.alertnet.org/thenews/newsdesk/HKG80770.htm (cited June 29, 2007).

National Biodiesel Board. "Bioheat frequently asked questions." (2007). Available from http://www.biodiesel.org/markets/hom/faqs.asp (cited July 29, 2007).

_____. "Biodiesel, next stop heating oil." (2007). Available from http://www.biodiesel.org/markets/hom/default.asp (cited July 29, 2007).

_____. "Standards and Warranties." (2007). Available from http://www.biodiesel.org/resources/fuelfactsheets/standards_and_warranties.shtm (cited July 30, 2007).

_____. "Engine Warranties." National Biodiesel Board (2007). Available from http://www.biodiesel.org/resources/fuelfactsheets/standards_and_warranties.shtm (cited July 30, 2007).

_____. "NBB Membership Information." (2007). Available from http://www.biodiesel.org/members/info/add_prod.shtm (cited July 29, 2007).

_____. "Biodiesel Production and Quality Fact Sheet." (2007). Available from http://www.biodiesel.org/pdf_files/fuelfactsheets/prod_quality.pdf (cited August 8, 2007).

_____. "Estimated US Biodiesel Sales." (2007). Available from http://www.biodiesel.org/pdf_files/fuelfactsheets/Biodiesel_Sales_Graph.pdf (cited August 8, 2007).

_____. "Emissions." (2007). Available from http://www.biodiesel.org/pdf_files/fuelfactsheets/emissions.pdf (cited August 8, 2007).

_____. "What Is Biodiesel?" (2007). Available from http://www.biodiesel.org/resources/definitions/ (cited August 22, 2007).

_____. "Biodiesel for Electrical Generation." (2007). Available from http://www.biodiesel.org/markets/ele/ (cited August 22, 2007).

_____. "How much biodiesel has been sold in the US?" (2007). Available from http://www.biodiesel.org/resources/faqs/ (cited September 26, 2007).

_____. "Fuel Quality Policy." In *Fuel Quality Enforcement Guide.* (2006).

_____. "Membership Packet." (August 18, 2006). Available from http://biodiesel.org/members/membershippacket/NBBMembershipPacket.pdf (cited July 29, 2007).

_____. "OEM Warranty Statements and Use of Biodiesel Blends over 5% (B5)." (2005). Available from http://nbb.org/pdf_files/B5_warranty_statement_32206.pdf (cited July 30, 2007).

_____. "EPA off-road low-sulfur rule could spur greater biodiesel use" [Press release]. (May 11, 2001). Available from http://www.biodiesel.org/resources/pressreleases/gen/20040511_EPA_Offroad_low_sulfur.pdf (cited February 1, 2005).

_____. "Summary Results from NBB/USEPA Tier 1 Health and Environmental Effects Testing for Biodiesel under the Requirements for USEPA Registration of Fuels and Fuel Additives (40CFR Part 79, Sec 21 1 (b)(2) and 21 1 (e))." (1998). Available from http://www.worldenergy.net/pdfs/TierIResults.pdf (cited August 8, 2007).

_____. "National Biodiesel Board Issue Brief: Biodiesel Tax Credit Implementation." (n.d.). Available from http://www.biodiesel.

org/news/taxincentive/Biodiesel%20Tax%20Credit%20NBB%20 Issue%20Breif.pdf (cited July 29, 2007).

National Biodiesel Board and AgriFuels. "Biodiesel Quality Awareness." Paper read at National Conference of Weights and Measures, July 9-13, 2006, at Chicago, Illinois.

National Cooperative Business Association. "How to start a cooperative." NCBA (n.d.). Available from http://www.ncba.coop/ abcoop_howto.cfm (cited July 27, 2007).

_____. "Co-op Principles and Values." (n.d.). Available from http:// www.ncba.coop/abcoop_ab_values.cfm (cited July 5, 2007).

National Renewable Energy Laboratory. "NREL B20 Study Shows No Increase in NOx Emissions." (November 9, 2006). Available from http://www.greencarcongress.com/2006/11/nrel_b20_ study_.html (cited July 29, 2007).

NC Division of Pollution Prevention and Environmental Assistance. "Oil and Grease Management." (1999). Available from www. p2pays.org/ref/04/03063.ppt (cited July 29, 2007).

Navarro, Xavier. "FOEI puts the blame on Wilmar for rainforest destruction caused by biodiesel production." (July 5, 2007). Available from http://www.autobloggreen.com/2007/07/05/foei-puts-the-blame-on-wilmar-for-rainforest-destruction-caused/ (cited September 4, 2007).

_____. "Biodiesel producers asking for certified oils." (July 18, 2007). Available from http://www.autobloggreen.com/2007/07/18/ biodiesel-producers-asking-for-certificated-oils/ (cited September 4, 2007).

Natural Resources Defense Council. "HECO and NRDC Finalize Biodiesel Purchase Policy." (2007). Available from http://www. nrdc.org/media/2007/070821.asp (cited September 4, 2007).

New Hampshire Department of Environmental Services. "Environmental Permitting: Regulations and Other Requirements Related to the Manufacture of Biodiesel." (2006). Available from http://www.des.state.nh.us/factsheets/co/co-16.htm (cited June 21, 2007).

News, BBC. "OPEC and China form closer ties." (December 22, 2005). Available from http://news.bbc.co.uk/2/hi/business/4551486.stm (cited September 27, 2006).

Nilles, Dave. "Combating the Glycerin Glut." *Biodiesel Magazine*, September 2006. Available from http://www.biodieselmagazine.com/article.jsp?article_id=1123 (cited June 5, 2007).

Oak Ridge National Laboratory and US Department of Agriculture. *Biomass as Feedstock for a Bioenergy and Bioproducts Industry: The Technical Feasibility of a Billion-Ton Annual Supply.* Washington, D.C.: US Department of Energy, Office of Energy Efficiency and Renewable Energy, Office of the Biomass Program, and the US Department of Agriculture, 2005.

Ohio Environmental Protection Agency. "Want to Start a Biodiesel Production Operation? Environmental Compliance Basics." (April 2007). Available from http://www.epa.state.oh.us/ocapp/sb/publications/biodieselguide.pdf (cited August 21, 2007).

Ohio Office of Compliance Assistance and Pollution Prevention. "Understanding the Spill Prevention, Control and Countermeasure (SPCC) Requirements." (2007). Available from http://www.epa.state.oh.us/ocapp/sb/publications/spcc.pdf (cited August 18, 2007).

Oil and Gas Journal. "Worldwide Look At Reserves and Production." *Oil and Gas Journal* 103 (47) (2005): 24-25.

_____. "Worldwide Look at Reserves and Production." *Oil and Gas Journal* 103 (47) (2005): 24-25.

Oregon Department of Environmental Quality. "Manufacturing Biodiesel: Questions and Answers for Oregonians Interested in Manufacture of Biodiesel for Personal Use or Small Scale Commercial Production." (September 21, 2006). Available from http://www.deq.state.or.us/aq/factsheets/06-AQ-014biodiesel.pdf (cited May 17, 2007).

Organization of Petroleum Exporting Countries. *Annual Statistical Bulletin 2004.* Vienna, Austria: OPEC, 2005.

_____. "About Us." (September 27, 2006). Available from http://www.opec.org/aboutus/ (cited July 29, 2007).

_____. "About Us." (2007). Available from http://www.opec.org/aboutus/ (cited July 29, 2007).

Pahl, Greg. *Biodiesel: Growing A New Energy Economy.* White River Junction, Vermont: Chelsea Green, 2005.

_____. *The Citizen Powered Energy Handbook: Community Solutions to a Global Crisis.* White River Junction, Vermont: Chelsea Green Publishing, 2007.

_____. "Community Supported Energy Offers a Third Way." (March 12, 2007). Available from http://www.renewableenergyaccess.com/rea/news/reinsider/story?id=47700 (cited June 27, 2007).

Pearl, G.G. "Animal Fat Potential for Bioenergy Use." Paper read at Bioenergy 2002: The Tenth Biennial Bioenergy Conference, September 22-26, 2002, at Boise, Idaho.

Peterson, Charles. "Potential Production of Biodiesel." University of Idaho, Department of Biological and Agricultural Engineering (n.d.). Available from http://www.uidaho.edu/bioenergy/BiodieselEd/publication/02.pdf (cited August 22, 2007).

Petit, J.R., J. Jouzei, D. Raynaud, N.I. Barkov, J.M. Barnola, I. Basile, M. Benders, J. Chappellaz, M. Davis, G. Delaygue, M. Delmotte, V.M. Kotlyakov, M. Legrand, V.Y. Lipenkov, C. Lorius, L. Peplin, C. Ritz, E. Saltzmani, and M. Stievenar. "Climate and atmospheric history of the past 420,000 years from the Vostok ice core, Antarctica." *Nature* 399 (1999): 429-436.

Pham, J. Peter. "China Goes on Safari." *World Defense Review* (August 2006). Available from http://worlddefensereview.com/pham082406.shtml (cited July 29, 2007).

Potter, Trent, and Don McCaffery. *Biodiesel in Australia - small scale production*. Griffith, NSW: Irrigation Research & Extension Committee, 2006.

Provost, Ken. "Ethanol-based Biodiesel. Journey to Forever" (n.d.). Available from http://journeytoforever.org/ethanol_link. html#ethylester (cited June 29, 2007).

Renewable Development Initiative. "Projects: Victoria Biodiesel." (March 28, 2006). Available from http://ebrdrenewables.com/ sites/renew/Lists/Projects/DispForm.aspx?ID=1762 (cited April 21, 2007).

Renewable Energy Access. "$30M Biodiesel Plant Slated for Argentina." RenewableEnergyAccess.com (January 20, 2006). Available from http://www.renewableenergyaccess.com/rea/news/ story?id=42102 (cited July 27, 2007).

Renewable Fuels Association. "Federal Regulations: Biodiesel Tax Credits." (2005). Available from http://www.ethanolrfa.org/ policy/regulations/federal/biodiesel/ (cited August 21, 2007).

Ring, Ed. "Biofuel or Biohazard?" (2007). Available from http:// www.sacramentoexecutive.com/2007/05/biofuel_or_biohaz- ard_by_ed_rin.html (cited June 29, 2007).

Roundtable on Sustainable Palm Oil (RSPO). "RSPO Principles and Criteria for Sustainable Palm Oil Production." (2006). Available from http://www.rspo.org/resource_centre/RSPO%20 Criteria%20Final%20Guidance%20with%20NI%20Document. pdf (cited September 4, 2007).

RNCOS. "Biofuel Market Worldwide (2006)." (November 2006). Available from http://www.rncos.com/Report/IM043.htm (cited July 29, 2007).

Schumacher, Joel. "Small Scale Biodiesel Production: An Overview." Agricultural Marketing Policy Paper No. 22, (May 2007). Available from http://www.ampc.montana.edu/policypaper/policy22. pdf (cited June 24, 2007).

Schumacher, Leon. "Biodiesel Lubricity." (2005). Available from http://www.uidaho.edu/bioenergy/BiodieselEd/publication/06.pdf (cited 2006).

Sheehan, John, Vince Camobreco, James Duffield, Michael Graboski, and Housein Shapouri. "Life cycle inventory of biodiesel and petroleum diesel for use in an urban bus: final report." Washington, D.C.: US Department of Agriculture and US Department of Energy (1998). Available from http://www.nrel.gov/docs/legosti/fy98/24089.pdf (cited August 8, 2007).

Sheehan, J., T. Dunahay, J. Benemann, and P. Roessler. "A Look Back at the US Department of Energy's Aquatic Species Program - Biodiesel from Algae." US DOE Office of Fuels Development and the National Renewable Energy Laboratory, 1998.

Shipper, L., and S. Peake. "Transport, Energy, and Climate Change." In *Energy and Environment Policy Analysis Series*. Paris, France: Organization for Economic Cooperation and Development, 1997.

_____. "Transport, Energy, and Climate Change." In *Energy and Environment Policy Analysis Series*. Paris, France: Organization for Economic Cooperation and Development, 1997.

Sine Nomine Group. "Biodiesel Distribuition Infrastructure." (January 26, 2006). Available from http://www.fleetchallenge.ca/en/library/presentations/biodiesel_distribution.pdf (cited July 12, 2006).

Singapore Government. "Budget 2007: Building Singapore into a Global Business Hub." (January 4, 2007). Available from http://www.sedb.com/edb/sg/en_uk/index/news_room/publications/singapore_investment3/singapore_investment2/budget_2007__building.html (cited June 29, 2007).

STA. "Slovenia builds its biggest biodiesel plant." B92 (January 29, 2007). Available from http://www.b92.net/eng/news/globe-article.php?yyyy=2007&mm=01&dd=29&nav_category=123&nav_id=39326 (cited April 21, 2007).

State of Ohio Environmental Protection Agency. "Want to Start a Biodiesel Production Operation? Environmental Compliance Basics." (April 2007). Available from http://www.epa.state.oh.us/ocapp/sb/publications/biodieselguide.pdf (cited June 22, 2007).

Steiman, Matthew. "Biodiesel Safety and Best Management Practices for Small-Scale Individual Use Production: See Title 25, PA Code, Chapter 245; Title 25, PA Code 299.122; and Title 25, PA Code, 285.122." (2007).

Stieger, W. "Sunfuel - The Way to Sustainable Mobility: Volkswagon AG Art." (2005).

Tans, Pieter P., and T.J. Conway. "Monthly Atmospheric CO2 Mixing Ratios from the NOAA CMDL Carbon Cycle Cooperative Global Air Sampling Network, 1968-2002." In *Trends: A Compendium of Data on Global Change. Carbon Dioxide Information Analysis Center.* Oak Ridge, Tennessee: Oak Ridge National Laboratory, 2005.

The Governor's Office of Regulatory Assistance, State of Washington. "Biodiesel Facility Permits Fact Sheet." (2006). Available from http://www.ora.wa.gov/documents/BiodieselFactSheet.pdf (cited July 27, 2007).

Thomas, Sharon, and Marcia Zalbowitz. "Fuel Cells - Green Power." Los Alamos National Laboratory (1997). Available from http://www.scied.science.doe.gov/nmsb/hydrogen/Guide%20to%20Fuel%20Cells.pdf (cited July 29, 2007).

U.S. Census Bureau. "U.S. and World Population Clocks - POPClocks." U.S. Census Bureau (July 29, 2007). Available from http://www.census.gov/main/www/popclock.html (cited July 29, 2007).

U.S. Central Intelligence Agency. "Unclassified Report to Congress on the Acquisition of Technology Relating to Weapons of Mass Destruction and Advanced Conventional Munitions." (2003). Available from https://www.cia.gov/cia/reports/721_reports/july_dec2003.htm#iraq (cited September 27, 2006).

US Department of Agriculture and US Department of Energy. "Life Cycle Inventory of Biodiesel and Petroleum Diesel for Use in an Urban Bus." (May 1998). Available from http://www.nrel.gov/docs/legosti/fy98/24089.pdf (cited July 29, 2007).

US Department of Agriculture, Economic Research Service. "China's Economic Growth Faces Challenges." US Department of Agriculture. (February 2005). Available from http://www.ers.usda.gov/AmberWaves/February05/Findings/ChinaEconomicGrowth.htm (cited September 27, 2006).

_____. "Amber Waves." US Department of Agriculture (2006). Available from http://www.ers.usda.gov/AmberWaves/September06/Indicators/Indicators.htm (cited July 29, 2007).

_____. "Data Sets." US Department of Agriculture (2006). Available from http://www.ers.usda.gov/Data/ (cited September 23, 2006).

US Department of Commerce, Bureau of Export Administration. "The effect on the national security of imports of crude oil and refined petroleum products." US Department of Commerce (November 1999). Available from http://efoia.bis.doc.gov/sec232/crudeoil/Sec232Oil1199.pdf.

US Department of Energy. "Alabama City Makes Fuel for Schools." (October 2004). Available from http://www.eere.energy.gov/state_energy_program/project_brief_detail.cfm/pb_id=757 (cited July 29, 2007).

_____. "Biodiesel Handling and Use Guidelines." US Department of Energy (2006). Available from http://www.nrel.gov/vehiclesandfuels/npbf/pdfs/40555.pdf (cited April 26, 2006).

_____. "U.S. Strategic Petroleum & other oil reserves, oil shale program." (July 17, 2007). Available from http://www.fossil.energy.gov/programs/reserves/npr/NPR_Oil_Shale_Program.html (cited July 29, 2007).

US Department of Energy, Energy Information Administration. "Sources of Oil Imports." (2003). Available from http://www. gravmag.com/oil.html#imports (cited July 29, 2007).

_____. "Annual Energy Review 2003." US Department of Energy (2003). Available from http://www.eia.doe.gov/emeu/aer/pdf/ pages/sec5.pdf (cited October 19, 2004).

_____. "Top World Oil Consumers." (2004). Available from http:// www.eia.doe.gov/emeu/cabs/topworldtables3_4.html (cited September 27, 2006).

_____. "U.S. Carbon Dioxide Emissions from Energy Sources: 2003 Flash Estimate." (June 2004). Available from http://www.eia.doe. gov/oiaf/1605/flash/flash.html (cited 2004).

_____. "Table 8.2: World Estimated Recoverable Coal." (2004). Available from http://www.eia.doe.gov/pub/international/ iea2005/table82.xls (cited July 29, 2007).

_____. "Annual Energy Outlook 2004." US Department of Energy (2004). Available from http://www.eia.doe.gov/neic/brochure/ aeo2004/aeo2004.htm (cited July 29, 2007).

_____. "Annual Energy Outlook 2004." U.S. Department of Energy (2004). Available from http://www.eia.doe.gov/neic/brochure/ aeo2004/aeo2004.htm (cited July 29, 2007).

_____. "Top World Oil Consumers, 2004." US Department of Energy (2004). Available from http://www.eia.doe.gov/emeu/cabs/ topworldtables3_4.html (cited September 27, 2006).

_____. "International energy outlook: oil resources in the 21st century." US Department of Energy (2004). Available from http://www.eia. doe.gov/oiaf/ieo/special_topics.html (cited December 2005).

_____. "Country Analysis Briefs." (November 2005). Available from http://www.eia.doe.gov/emeu/cabs/usa.html (cited July 29, 2007).

_____. "E85 Fleet Toolkit." US Department of Energy (March 8, 2005). Available from http://eeredev.nrel.gov/afdc/e85toolkit/ fueling_options.html (cited July 12, 2006).

_____. "International Energy Outlook 2006." US Department of Energy (2006). Available from http://www.eia.doe.gov/oiaf/ieo/pdf/ieoreftab_4.pdf (cited July 27, 2007).

_____. "Country Analysis Briefs: China." (August 2006). Available from http://www.eia.doe.gov/emeu/cabs/China/Background.html (cited September 27, 2006).

_____. "International Energy Outlook 2006." (2006). Available from http://www.eia.doe.gov/oiaf/ieo/world.html (cited September 27, 2006).

_____. "Form EIA-820, Annual Refinery Report." (January 2006). Available from http://www.eia.doe.gov/pub/oil_gas/petroleum/data_publications/refinery_capacity_data/current/table1.pdf (cited July 10, 2006).

_____. "A Primer on Gasoline Prices." US Department of Energy (2006). Available from http://www.eia.doe.gov/bookshelf/brochures/gasolinepricesprimer/eia1_2005primerM.html (cited July29, 2007).

_____. "U.S. Crude Oil Imports by Country of Origin." (June 21, 2007). Available from http://tonto.eia.doe.gov/dnav/pet/pet_move_impcus_a2_nus_ep00_im0_mbbl_m.htm (cited July 29, 2007).

_____. "Monthly Energy Review - Table 2.1." Washington, D.C. (2007).

_____. "Annual Energy Outlook 2007." US Department of Energy (February 2007). Available from http://www.eia.doe.gov/oiaf/aeo/pdf/0383(2007).pdf (cited July 26, 2007).

U.S. Environmental Protection Agency. " Purpose and Applicability of Speculative Accumulation Provision." (1995). Available from http://yosemite.epa.gov/osw/rcra.nsf/ea6e50d-c6214725285256bf00063269d/494fb09b855aa2c28525670f006bcdbf!OpenDocument (cited August 20, 2007).

_____. "The Emergency Planning and Community Right-to-Know Act." EPA 550-F-00-004 (2000). Available from http://yosemite. epa.gov/oswer/CeppoWeb.nsf/vwResourcesByFilename/epcra. pdf/$File/epcra.pdf (cited August 20, 2007).

_____. "Methanol." (2000). Available from http://www.epa.gov/ttn/ atw/hlthef/methanol.html (cited August 22, 2007).

_____. *Control of Air Pollution from New Motor Vehicles: Heavy-Duty Engine and Vehicle Standards and Highway Diesel Fuel Sulfur Control Requirements; Final Rule.* Washington, D.C.: Federal Register, 2001.

_____. "Inventory of US greenhouse gas emissions and sinks: 1990-2002." (April 2004). Available from http://yosem-ite.epa.gov/oar/globalwarming.nsf/UniqueKeyLookup/ RAMR5WNMKA/$File/04trends.pdf (cited October 13, 2004).

_____. "40 CFR 261.2, Table 1." (2005). Available from http://www. access.gpo.gov/nara/cfr/waisidx_05/40cfr261_05.html (cited August 19, 2007).

_____. "Spill Prevention, Control and Countermeasure." US EPA Oil Program (December 28, 2006). Available from http://www. epa.gov/oilspill/spcc.htm (cited June 24, 2007).

_____. "About Air Toxics, Health and Ecological Effects." (March 7, 2007). Available from http://www.epa.gov/air/toxicair/newtoxics. html (cited July 29, 2007).

_____. "What are the Six Common Air Pollutants?" (July 23, 2007). Available from http://epa.gov/air/urbanair/ (cited July 29, 2007).

U.S. Government Printing Office. "Registration of Fuels and Fuel Additives." (2001). Available from http://www.access.gpo.gov/nara/ cfr/waisidx_01/40cfr79_01.html (cited February 10, 2005).

Underwriters Laboratories Inc. "Steel Aboveground Tanks for Flammable and Combustible Liquids: UL 142." Underwriters

Laboratories Inc. (2007). Available from http://ulstandardsinfonet.ul.com/scopes/scopes.asp?fn=0142.html (cited May 26, 2007).

United States Department of Energy, Energy Information Association. "Annual Energy Outlook." (2004).

University of Idaho. "Moisture Absorption in Biodiesel." *Biodiesel TechNotes*, Spring, 2 (2006).

Urbanchuk, John M. "Contribution of the Ethanol Industry." Renewable Fuels Association (2006). Available from http://www.ethanol.org/documents/EthanolEconomicContributionFeb06.pdf (cited July 11, 2006).

Valente, Marcela. "Agriculture – Argentina: Soy Overruns Everything in Its Path." IPS (August 6, 2006). Available from http://www.ipsnews.net/interna.asp?idnews=24977 (cited July 27, 2007).

_____. "Argentina: The Environmental Costs of Biofuel." IPS (April 20, 2006). Available from http://ipsnews.net/news.asp?idnews=32959 (cited July 27, 2007).

Vanderbosch, R.H. "Bio-oil Applications." Biomass Technology Group (March 2003). Available from http://www.btgworld.com/technologies/bio-oil-applications.html (cited September 27, 2006).

Virginia Department of Environmental Quality. Personal communication. VADEQ Regulations, August 2007.

Wald, M.L. "Both Promise and Problems for New Tigers in Your Tank." *New York Times*, October 26, 2005, 1.

Wang, M., C. Saricks, and D. Santini. *Effects of Fuel Ethanol Use on Fuel-Cycle Energy and Greenhouse Gas Emissions, ANL/ESD-38*. Argonne, Illinois: Argonne National Laboratory, Center for Transportation Research, 1999.

Waterloo School for Community Development. http://en.wikipedia.org/wiki/Biodiesel_around_the_World "The Working Centre." (n.d.). Available from http://www.theworkingcentre.org/wscd/ctx/story/story.html (cited July 27, 2007).

Wedel, R. "Technical Handbook for Marine Biodiesel in Recreational Boats." In *CytoCulture*. National Renewable Energy Laboratory, 1999.

Wikipedia. "Biodiesel around the world." (July 6, 2007). Available from http://en.wikipedia.org/wiki/Biodiesel_around_the_World (cited February 16, 2007).

_____. "Statement on the Co-operative Identity." (July 5, 2007). Available from http://en.wikipedia.org/wiki/Statement_on_the_ Co-operative_Identity (cited July 29, 2007).

Wiltsee, G. "Urban waste grease resource assessment: Appel Consultants, Inc." NREL, 1998.

Wired News. "Automakers Give Biodiesel a Boost." *Wired News* (September 23, 2004). Available from http://www.wired. com/news/autotech/0,2554,65054,00.html (cited July 30, 2007).

Wisconsin Department of Commerce. "Biodiesel: General Production Regulations in Wisconsin." (n.d.). Available from http:// www.uwex.edu/ces/ag/teams/grains/documents/Biodiesel_GeneralProductionReguationsinWisconsin.pdf (cited November 26, 2006).

World Energy Council. "Natural Bitumen and Extra Heavy Oil: WEC Survey of Energy Resources." (2001). Available from http://www.worldenergy.org/wec-geis/publications/reports/ser/ bitumen/bitumen.asp (cited September 27, 2006).

World Meteorological Association, Global Atmosphere Watch. "WMO Greenhouse Gas Bulletin: The State of Greenhouse Gases in the Atmosphere Using Global Observations up to December 2004: United Nations World Meteorological Organization, Environment Division, Atmospheric Research and Environment Programme." (2005).

World Oil. "Estimated Proven World Reserves." *World Oil* 226 (9) (2005).

World Rainforest Movement. "World Rainforest Movement." *World Rainforest Bulletin* 112 (November 2006). Available from http://www.wrm.org.uy/bulletin/112/Colombia.html (cited June 29 2007).

Yang, Y., and J. Strozeski. *Politics in Africa and American Security Interests.* Harrisonburg, Virginia: James Madison University, 2006.

Bios

Amy Townsend, Ph.D.

Amy Townsend is Founder and President of Sustainable Development International Corporation (SDIC) and The Hungry Vegan. Specializing in the areas of green business and green building, she has worked in the US and abroad with governments and organizations interested in becoming more environmentally friendly. She has written three books on green business as well as a weekly newpaper column and several magazine articles on greener living.

Dr. Townsend is an adjunct assistant professor in James Madison University's Department of Integrated Science and Technology where she became involved with biodiesel. She has worked with Malta's largest biodiesel producer to determine the feasibility of collecting household waste oil and, later, to find ways to enhance the house waste oil collection scheme.

Dr. Townsend also is an adjunct professor of Marlboro College's Sustainable MBA program and is a member of the US Green Building Council.

Billy Broas

With a background in biotechnology and alternative fuels, Billy Broas has been involved in several biodiesel projects. As part of a team, he worked with Malta's largest biodiesel producer to develop a marketing strategy to increase household participation in waste oil collection. Additionally, Mr. Broas investigated the technological and economic feasibility of producing biodiesel fuel at James Madison University using waste cooking oil produced at campus dining halls and using the fuel to offset petroleum diesel fuel consumption for university vehicles.

As an energy consultant for the Antares Group Inc., he works with clients on renewable energy projects, including waste-to-energy plants, biomass fuel supply studies, alternative fuels, and others. As a craft beer lover, he has developed a social networking web site for home brewers and other craft beer enthusiasts (www.FriendlyBrew.com).

Chelsea Jenkins

Chelsea Jenkins is Coordinator for the Hampton Roads Clean Cities Coalition. She leads technical, awareness, and funding initiatives promoting the use of alternative fuels for transportation. Her biodiesel journey began at James Madison University, in Virginia, working as an undergraduate assistant in the alternative fuels program within the Department of Integrated Science and Technology. After becoming oddly infatuated with this renewable alternative to petroleum diesel fuel, her biodiesel adventures led her abroad to conduct a feasibility study for the largest biodiesel producer in Malta. Malta is where Chelsea met Dr. Townsend. She has since worked on alternative fuels, energy, air quality, and sustainability projects in the U.S. and abroad.

Kevin Ray

Kevin Ray's involvement with biodiesel has focused on energy, engineering and manufacturing, and business and technology. Much of his biodiesel research has focused on biodiesel regulators and small-scale producers in Virginia, culminating in the paper *Permitting and Regulatory Guidelines for Small-Scale Biodiesel Production in the Commonwealth of VA*. Additionally, Mr. Ray worked with a team in Malta to identify ways to improve an waste oil collection. Currently, he is working for IBM as a consultant.

Notes

Notes